RELATIVISTIC NUCLEAR PHYSICS IN THE LIGHT FRONT FORMALISM

RELATIVISTIC NUCLEAR PHYSICS IN THE LIGHT FRONT FORMALISM

V. R. Garsevanishvili
Mathematical Institute of the Georgian Academy of Sciences, Tbilisi, Republic of Georgia

and

Z.R. Menteshashvili
Institute of High Energy Physics, Tbilisi State University, Tbilisi, Republic of Georgia

NOVA SCIENCE PUBLISHERS, INC.

Nova Science Publishers, Inc.
6080 Jericho Turnpike, Suite 207
Commack, New York 11725

Book Production Manager: Janet Glanzman White
Graphics: Elenor Kallberg, Christopher Concannon

Library of Congress Cataloging-in-Publication Data available upon request

ISBN 1-56072-093-X

Printed in the United States of America

Contents

Preface vii

Introduction 1

I. Light Front Formalism for Bound States 9
   1. Equation for the Two-Body Bound State Wave Function 9
   2. Equation for the Scattering Amplitude and Relation to the Equation in the Infinite Momentum Frame 13
   3. Equation for the Many-Body Bound State Wave Function 15

II. Relativistic Elastic Form Factors and Scattering Amplitudes for Composite Systems 21
   1. Formulation of the Method 21
   2. Elastic Form Factors in the Impulse Approximation 25
   3. Relativistic Form Factors for the Many-Body System 27
   4. Scattering of Relativistic Composite Systems 32
   5. Gluon Exchange in the Intermediate State 35
   6. Constituent Interchange Mechanism 37

III. Deep Inelastic Form Factors of Composite Systems and Multiquark States in Nuclei 41
   1. Construction of Tensor $W_{\mu\nu}$ 42
   2. Lowest Order in the Electromagnetic Interaction 48
   3. Model Parametrizations of Wave Functions 54
   4. Quark Degrees of Freedom in Nuclei and EMC-Effect 57

IV. Processes Involving High Energy Nuclei and The Problem of Relativization of Nuclear Wave Functions 71
   1. Scale-Invariant Parametrization of the Deuteron Relativistic Wave Function 71
   2. Break-Up of the Relativistic Deuteron and the Verification of Scale Invariant Properties of Its Wave Function 74
   3. The Break Up of More Complicated Nuclei 86
   4. The Scaling Properties of the Relativistic Wave Functions of ^{4}He and ^{3}He Nuclei 97

V. Deep Inelastic Lepton-Nucleus Scattering with the Fragment Nucleus in the Final State 107
1. Deep Inelastic Scattering of Charged Lepton on Nucleus, The Process 1A → 1'(A – 1)X 107
2. Deep Inelastic Scattering of the Neutrino (Antineutrino) on the Nucleus. The Process νA → 1(A – 1)X 114
3. Lepton Pair Production in Hadron-Nucleus Collisions. The Process aA → 1^+1^-(A – 1)X 119
4. Dimensional Analysis and the Principle of Automodelity 123
5. Relations Between the Nucleus and Nucleon Structure Functions 129

References 137

Subject Index 145

PREFACE

The book is devoted to the subject intensively developed in the last two decades, which is usually called relativistic nuclear physics. The experiments in this direction have been performed first in early 70-ies in Dubna and Berkeley with beams of nuclei with energies 2 - 10 GeV/nucleon. Now the experiments with beam and target nuclei are performed at CERN and other laboratories with beams of energies of hundreds of GeV/nucleon. In this energy range one hopes to find a new state of nuclear matter - quark-gluon plasma.

The book contains the results on the light front formalism for the composite systems and its applications to the problems of relativistic nuclear physics (elastic and deep inelastic scattering of leptons by nuclei, break-up processes of light nuclei etc.).

Let us list briefly the main topics, which are treated in the book:

The light front formalism for two and many body relativistic systems is developed. The equations of this approach can serve as a good tool for the treatment of a wide class of elementary particle and nuclear physics problems.

The problems of interactions of relativistic composite systems are considered. It is shown that the constituent interchange mechanism together with the power-law fall-off of quark distribution functions in hadrons leads to the results which are very close to the quark counting rules.

A method for constructing of elastic and deep inelastic formfactors and scattering amplitudes of composite particles is developed. It is shown that the taking into account of transverse motion of quarks leads to the violation of Bjorken scaling.

The detailed analysis of the so called EMC-effect is given. The explanation of the EMC-ratio is based on the scattering of electrons on the colourless multiquark states in nuclei.

The problem of relativisation of the nuclear wave functions is studied. The relativistic analogues of the well-known wave functions of light nuclei are obtained. These wave functions posess the scaling properties with respect to the light front variables in terms of which the longitudinal motion of constituents is parametrized.

The main points of the relativistic approach to the processes involving atomic nuclei are formulated. In particular, the calculation of some high energy processes involving nuclei is given. The scaling properties of observable quantities are predicted and checked by comparison with experiment.

The relativistic approach to the deep inelastic lepton-nucleus scattering and lepton pair production in hadron-nucleus interactions is developed when the final lepton and fragment of incoming nucleus are detected in coincidence. The kinematical analysis of these processes is given. The scaling properties of the corresponding structure functions are established.

The book will be interesting for the scientists and postgraduate students which are working in the field of high energy elementary particle and nuclear physics.

When working on the problems discussed in this book we had numerous helpful discussions with N.S.Amaglobeli, A.M.Baldin, N.N.Bogolubov, V.G.Kadyshevsky, T.I.Kopaleishvili, A.A.Logunov, V.A.Matveev, R.G.Salukvadze, A.N.Tavkhelidze for whom we express our sincere thanks. Some of the problems considered here we discussed with S.J.Brodsky, M.Chaichian, M.A.Eliashvili, S.B.Gerasimov, A.A.Khelashvili, H.Leutwyler, R.M.Mir-Kasimov, A.N.Mitra, E.Predazzi, R.Rosenfelder, M.D.Scadron, L.A.Slepchenko, F.G.Tkebuchava. The experimental situation was discussed with the members of experimental groups of A.Fridman, V.V.Glagolev, M.S.Nioradze, G.Roche. Some of the results presented in the book have been obtained in collaboration with Ya.Z.Darbaidze, Sh.M.Esakia, R.N.Faustov, D.G.Mirianashvili. To all of them we express our deep gratitude.

INTRODUCTION

The study of high-energy particle-nucleus interactions is one of the basic tasks of elementary particle and nuclear physics. At different stages this task was posed and solved both experimentally and theoretically depending on the available experimental facilities and theoretical concepts which for several decades had been based on the quantum field theory.

Recent research has given an insight into the complicated relativistic structure of hadrons. Besides, the study of high-energy nucleus interactions has shown that idea of nuclei as systems of quasi-independent nonrelativistic nucleons is incomplete. Hence it has became necessary to take into account the relativistic character of intrinsic motion of nucleons and to consider the quark degrees of freedom in nuclei.

The above-stated reasons have motivated the subject of research considered in this book.

According to modern concepts, nucleons and nuclei are bound states of more elementary constituents. Interactions between them or their interaction with particles which at the present stage can be considered as elementary are principal sources of information on bound states. Since at high energies inelastic processes play the dominant role (they account for up to 80 % of all events), to reconstruct the complete picture of the final state is a practically unrealizable task both from theoretical and experimental standpoint. That is why most of recent experiments in high-energy physics are inclusive.

Investigations with beams of high-energy nuclei carried out in Dubna,

Berkeley and other laboratories since the begining of the seventies have stimulated interest in the study of nuclei as relativistic composite systems and the processes of interaction between them (see, e.g. review papers [1 - 29] and the references therein). Presently, there are beams of high-energy nuclei with various atomic number in a wide energy range from several GeV/nucleon up to several hundred GeV/nucleon at CERN [26, 28, 30-32]. Note that this range corresponds to relativistic physics and the non-relativistic theory of nuclear reactions (see, e.g., the review paper [33,34] and references therein) based on the Schroedinger formalism of quantum mechanics does not work in this case. Relativistic effects have to be taken into account in this energy region and high-energy phenomena can be adequately described only using the relativistically invariant formalism. Thus it seems more convenient to plot also the data in terms of variables which have the relativistic origin. Investigations in the field of relativistic nuclear physics have been stimulated by the prediction [35] of the cumulative particle production in the nucleus-nucleus interactions and by the experimental observation of this effect [36].

In the majority of experiments involving beams of relativistic nuclei the inclusive spectra of secondary particles (pions, protons, light nuclei) have been measured in a wide range of kinematic variables and for various beam and target nuclei. The energy of a secondary particle can be related to the cumulative number N_{min} which is the effective number of nucleons of the fragmenting nucleus involved in the process. This variable is a complicated combination of the momentum and the emission angle of the outgoing particle and in the high energy limit it takes a simple form [8, 15, 18]

$$N_{min} = \begin{cases} \dfrac{E - p_z}{m} & \text{in the beam rest frame} \\ \dfrac{E + p_z}{E^o + p_z^o} & \text{in the target rest frame} \end{cases}$$

Here E and p_z are the energy and the z-component of momentum of the secondary particle, m is the proton mass, E^a and p^a are the energy and the z-component of the momentum per nucleon of the incident nucleus. The variable N_{min} is closely related to the light front variables which will be discussed further on.

A simple exponential dependence of the inclusive cross section on the variable N_{min}

$$E\frac{d\sigma}{d\vec{p}} \sim C\exp(-aN_{min})$$

(the parameters a and C do not depend on the properties of the target in the beam fragmentation region) describes satisfactorily momentum and angular distributions of secondary particles. The region of cumulative production corresponds to $N_{min} > 1$. This means that the production occurs as a result of interaction with a group of nucleons of the fragmenting nucleus.

Various theoretical schemes have been suggested to describe processes involving relativistic nuclei, but no model can explain all the experimental data in a self-consistent way.

In the experiments with beams of high-energy nuclei the situations often occur when small internucleon distances must be taken into account. This corresponds to large relative momenta of constituents of the nucleus. The problem arises to describe adequately nuclei at arbitrary momenta of its constituents. That is why it is more convenient to describe the processes involving relativistic nuclei in terms of the wave functions, in which the relativistic character of nucleon motion is taken into account, instead of the ordinary quantum-mechanical nuclear wave functions, which correspond to the motion of nucleons with small internal momenta.

This brings us back to the question of bound states and we recall that in the quantum field theory the two-particle system can be described by means of the Bethe-Salpeter formalism [37]. However, the dependence of the Bethe-Salpeter wave function on the relative time of two particles makes it difficult the probabilistic interpretation of this wave function and gives rise to a number of mathematical problems. On the other hand, the relativistic three-dimensional approach of Logunov-Tavkhelidze [38] based on the two-time formalism for Green functions enables one to avoid the difficulty connected with the relative time and is much simpler in applications.

We give here the derivation of the three-dimensional equations in the light front formalism. This form of the three-dimensional approach turned out to be effective in the treatment of high energy interactions of composite systems. In particular, it gives a number of scale invariant predictions for observable quantities which can be checked experimentally.

In this approach the relativistic nucleus with the total 4-momentum P is described by means of the relativistic wave function $\Phi_P([x_i,\vec{p}_{i,\perp}])$ [7,17,25,39], where the "longitudinal motion" of constituents is parametrized by means of the scale-invariant variables

$$x_i = \frac{p_{i,0}+p_{i,z}}{P_0+P_z}$$

where $p_{i,\mu}$($\mu=0,1,2,3$ is the Lorentz index) and P_μ are an individual 4-momentum of the i-th constituent and the total momentum of the system, respectively. Variables x_i are ratios of the light front variables. In terms of these variables the wave function of the composite system reflects, in particular, the dependence of the internal motion of constituents on the total momentum of the system.

In what follows upper-case letters will denote the characteristics (momenta, masses, etc.) of the composite systems and lower-case letters the characteristics of constituents. Square brackets in the argument of the wave function Φ_P denote the set of the variables x_i and $\vec{p}_{i,\perp}$ which satisfy the condition:

$$\sum_{i=1}^{A} x_i = 1; \quad 0 < x_i < 1; \quad \sum_{i=1}^{A} \vec{p}_{i,\perp} = \vec{P}_\perp$$

The book is organized as follows:

Chapter I is devoted to the formulation of the three-dimensional formalism for composite systems in terms of light front variables. Equations are derived for bound states and scattering problems. It is shown how equations of this approach are related with or differ from the Weinberg equation obtained in the framework of the old-fashioned perturbation theory in the infinite momentum frame.

Chapter II deals with the method of constructing relativistic elastic form-factors and scattering amplitudes of composite systems in the light front formalism. A general expression is obtained for the matrix element of the current of composite system in terms of relativistic wave functions and the generalized vertex operator $\tilde{\Gamma}_\mu$. The explicit form of the vertex operator is found in the impulse approximation and the electromagnetic form-factor is calculated for a system, consisting of two or arbitrary number of constituents.

Problems of the interaction of relativistic composite systems are also discussed in this chapter. The scattering amplitude is expressed in general form, using relativistic wave functions and the transition operator. The constituent interchange mechanism is considered.

Chapter III is devoted to the study of deep inelastic form-factors of composite systems. Like the case of elastic form-factors, the general expression for the deep inelastic tensor $W_{\mu\nu}$ is calculated in terms of the relativistic wave functions and the generalized two-photon vertex $\tilde{\Gamma}_{\mu\nu}$. The explicit form of this operator is calculated in the lowest order in the electromagnetic interaction and expressions are given for the structure functions W_1 and νW_2. It is shown that if the transverse motion of quarks is taken into account, the Bjorken scaling is violated and the structure functions become the square of the momentum transfer dependent. The problem of explaining of the so-called EMC-effect is stated. It is shown that one possible way to explain this effect is the scattering of electrons on the colourless multiquark configurations in nuclei.

In the Chapter IV problems of the relativization of nuclear wave functions are posed and solved. The case of the simplest nucleus - deuteron is considered in detail. A way to the relativization of the well-known nonrelativistic wave functions of the deuteron in indicated. The scaling properties of the relativistic wave function of the deuteron are checked by studying spectator-nucleon distributions in the deuteron break-up at various incident energies. It is shown that the scaling properties of the relativistic wave function hold with good accuracy in a wide energy range. Calculations are performed using the relativistic analogue of the well-known Hulten wave function of the deuteron.

Problems of relativization of wave functions of more complex nuclei are also considered. The differential cross section of the break-up of the relativistic nucleus on the hydrogen target is calculated. Theoretical calculations are compared with experimental data on the $^4He \to {}^3He$, $^3He \to d$ fragmentation. The analysis of data shows that the scaling properties of the relativistic wave functions of the ^4He and ^3He nuclei hold with a good accuracy in the energy range considered.

In the Chapter V deep inelastic lepton-nucleus processes in coinsidence are considered. The kinematics of these processes is studied in detail. The dimensional analysis of structure functions is performed and their scaling properties are established. It is shown that the structure functions of deep inelastic lepton-nucleus processes in coinsidence are related to the well-known structure functions of deep inelastic processes on the nucleon if it is assumed that the process occurs on the individual nucleons of the nucleus. Like in the preceding chapter, here also the overlap integral of relativistic wave functions of the initial and the spectator nuclei occurs. The lack of experimental data, however, makes it impossible for the time being to compare theoretical results with experiment.

Chapter I

LIGHT FRONT FORMALISM FOR BOUND STATES

1. Equation for the Two-Body Bound State Wave Function

In this section we develop the three-dimensional formalism in the light front variables. This variables were introduced by Dirac [40] with the aim to construct the quantum theory with canonical commutation relations on the light front hyperplane (instead of traditionally used $t=0$ hyperplane). We note that the real progress in this direction has been achieved much later [41-45].

Consider the Bethe-Salpeter amplitude (wave function)

$$\chi_{P,\alpha}=\langle 0|T\big(\varphi_1(x_1)\varphi_2(x_2)\big)|P,\alpha\rangle=e^{-iPX}\chi_{P,\alpha}(x) \qquad (1.1)$$

Here $|P,\alpha\rangle$ - is the state vector of two particles with total 4-momentum P and quantum numbers α, $X=(x_1+x_2)/2$ is the centre-of-mass coordinate, $P=p_1+p_2$. Define the relative coordinate and momentum

$$x=x_1-x_2\,, \qquad p=\frac{p_1-p_2}{2}\,, \qquad (1.2)$$

and introduce the light front variables

$$x_\pm=\frac{x_0\pm x_3}{2}\,, \quad p_\pm=p_0\pm p_3\,, \quad P_\pm=P_0\pm P_3\,. \qquad (1.3)$$

Introduce the Fourier transform $\chi_{P,\alpha}(p)=\chi_{P,\alpha}(p_-,p_+,\vec{p}_\perp)$ of the Bethe-Salpeter amplitude

$$\chi_{P,\alpha}(x)=\chi_{P,\alpha}(x_+,x_-,\vec{x}_\perp)=\int d^4p\, e^{-ipx}\chi_{P,\alpha}(p)=$$
$$=\frac{1}{2}\int dp_+dp_-d\vec{p}_\perp e^{-i(p_+x_-+p_-x_+-\vec{p}_\perp\vec{x}_\perp)}\chi_{P,\alpha}(p) \tag{1.4}$$

and define the function

$$\Psi_{P,\alpha}(p_+,\vec{p}_\perp)=\int_{-\infty}^{\infty}dp_-\,\chi_{P,\alpha}(p_-,p_+,\vec{p}_\perp) \tag{1.5}$$

It can be shown that the function $\Psi_{P,\alpha}(p_+,\vec{p}_\perp)$ depends on the values of the Bethe-Salpeter amplitude on the light front hyperplane

$$x_0+x_3=0$$

In fact, using definition (1.5) and Fourier transformation (1.4) we get:

$$\Psi_{P,\alpha}(p_+,\vec{p}_\perp)=\frac{2}{(2\pi)^3}\int dx_+dx_-d\vec{x}_\perp\delta(x_+)e^{-i(p_+x_--\vec{p}_\perp\vec{x}_\perp)}$$
$$\chi_{P,\alpha}(x_+,x_-,\vec{x}_\perp) \tag{1.6}$$

Consider now the two-particle Green function

$$G(x_1,x_2;x_1',x_2')=G(X-X';x,x')=$$
$$=\langle 0|T\big(\varphi_1(x_1)\varphi_2(x_2)\varphi_1^+(x_1')\varphi_2^+(x_2')\big)|0\rangle= \tag{1.7}$$
$$=\frac{1}{(2\pi)^8}\int dP\,dp\,dp'e^{-iP(X-X')-i(px-p'x')}G(P;p,p')$$

Here the total and relative 4-momenta and 4-coordinats in the initial and final states are introduced

$$P=p_1+p_2,\quad p=\frac{p_1-p_2}{2},\qquad X=\frac{x_1+x_2}{2},\quad x=x_1-x_2. \tag{1.8}$$

$$P'=p_1'+p_2',\quad p'=\frac{p_1'-p_2'}{2},\qquad X'=\frac{x_1'+x_2'}{2},\quad x'=x_1'-x_2'. \tag{1.9}$$

Define the Fourier transform of the "two-time" Green function

$$\tilde{G}(P;p_+,\vec{p}_\perp;p_+',\vec{p}_\perp')=\int_{-\infty}^{\infty}dp_-dp_-'G(P;p,p') \tag{1.10}$$

For free particles we have

$$G^{(o)}(P;p,p')=\frac{-\delta^{(4)}(p-p')}{\left[\left(\frac{P}{2}+p\right)^2-m_1^2+i\varepsilon\right]\left[\left(\frac{P}{2}-p\right)^2-m_2^2+i\varepsilon\right]} \tag{1.11}$$

Performing the integration according to the definition (1.10), we obtain [46]

$$\begin{aligned}&\tilde{G}^{(o)}(P;p_+,\vec{p}_\perp;p_+',\vec{p}_\perp')=\\ &=\frac{4\pi i\,\delta(p_+-p_+')\delta^{(2)}(\vec{p}_\perp-\vec{p}_\perp')\Theta(x)\Theta(1-x)}{P_+x(1-x)\left[P^2+\vec{P}_\perp^2-\frac{(\vec{P}/2+\vec{p})_\perp^2+m_1^2}{x}-\frac{(\vec{P}/2-\vec{p})_\perp^2-m_2^2}{1-x}\right]}=\\ &=\tilde{G}^{(o)}(P;p_+,\vec{p}_\perp)\delta(p_+-p_+')\delta^{(2)}(\vec{p}_\perp-\vec{p}_\perp')\end{aligned} \tag{1.12}$$

In this expression the variable x is introduced in the following way

$$x=\frac{1}{2}+\frac{p_+}{P_+} \tag{1.13}$$

It is obvious that when the variable x varies in the limits

$$0<x<1 \tag{1.14}$$

the variable p_+ varies in the interval $[-P_+/2\ ,\ P_+/2]$.

Define now the inverse operator by the relation

$$\int_{-P_+/2}^{P_+/2} dp''_+ \int d\vec{p}''_\perp \, \tilde{G}^{-1}(P;p_+,\vec{p}_\perp;p''_+,\vec{p}''_\perp)\,\tilde{G}(P;p''_+,\vec{p}''_\perp;p'_+,\vec{p}'_\perp) = \delta(p_+-p'_+)\,\delta^{(2)}(\vec{p}_\perp-\vec{p}'_\perp) \tag{1.15}$$

Introduce the interaction kernel V **(quasipotential):**

$$\tilde{G}^{-1}(P;p_+,\vec{p}_\perp;p'_+,\vec{p}'_\perp) = \tilde{G}^{(0)-1}(P;p_+,\vec{p}_\perp)\,\delta(p_+-p'_+)\,\delta^{(2)}(\vec{p}_\perp-\vec{p}'_\perp) - \frac{1}{4\pi i} V(P;p_+,\vec{p}_\perp;p'_+,\vec{p}'_\perp) \tag{1.16}$$

After simple transformations the equation for the wave function

$$\Phi_{P,\alpha}(x,\vec{p}_\perp) = P_+ x(1-x)\,\Psi_{P,\alpha}(p_+,\vec{p}_\perp) \tag{1.17}$$

takes the form

$$\left[P^2 - \frac{(\vec{p}_\perp+(1/2-x)\vec{P}_\perp)^2+m_1^2}{x} - \frac{(\vec{p}_\perp+(1/2-x)\vec{P}_\perp)^2+m_2^2}{1-x}\right]\Phi_{P,\alpha}(x,\vec{p}_\perp) = \int_0^1 \frac{dx'}{x'(1-x')}\int d\vec{p}'_\perp \, V(P;p_+,\vec{p}_\perp;p'_+,\vec{p}'_\perp)\,\Phi_{P,\alpha}(x',\vec{p}'_\perp) \tag{1.18}$$

The equation obtained gives the wave function of a bound state in an arbitrary Lorentz reference frame. Comparing it with the equation in the frame where $\vec{P}_\perp=0$ **we get the transformation property for the wave function from the arbitrary frame to the frame, in which the total transverse momentum of two-particle bound state is equal to zero:**

$$\Phi_P(x,\vec{p}_\perp) - \Phi_{\vec{P}_\perp=0}(x,\vec{p}_\perp+(1/2-x)\vec{P}_\perp) \tag{1.19}$$

For the sake of simplicity we do not consider here the spin case, which

is treated in detail in Ref. [47].

2. Equation for the Scattering Amplitude and Relation to the Equation in the Infinite Momentum Frame

Derive now the equation for the two-body scattering amplitude. The definition of the scattering amplitude $T(P;p,p')$ in the 4-dimensional covariant Bethe-Salpeter formalism looks as follows:

$$\begin{aligned} G(P;p,p') &= G^{(o)}(P;p,p') + \\ &+\int d^4p'' d^4p''' G^{(o)}(P;p,p'')T(P;p'',p''')G^{(o)}(P;p''',p') = \\ &= G^{(o)}(P;p)\delta^{(4)}(p-p') + G^{(o)}(P;p)T(P;p,p')G^{(o)}(P;p') \end{aligned} \tag{1.20}$$

Define the quantity $\tilde{T}(P;p_+,\vec{p}_\perp;p'_+,\vec{p}'_\perp)$ by the similar expression

$$\begin{aligned} \tilde{G}(P;p_+,\vec{p}_\perp;p'_+,\vec{p}'_\perp) &= \tilde{G}^{(o)}(P;p_+,\vec{p}_\perp)\delta(p_+-p'_+)\delta^{(2)}(\vec{p}_\perp-\vec{p}'_\perp) + \\ &+ \tilde{G}^{(o)}(P;p_+,\vec{p}_\perp)\tilde{T}(P;p_+,\vec{p}_\perp;p'_+,\vec{p}'_\perp)\tilde{G}^{(o)}(P;p'_+,\vec{p}'_\perp) \end{aligned} \tag{1.21}$$

Integrating (1.20) according to (1.10) we get:

$$\tilde{G} = \tilde{G}^{(o)} + \widetilde{G^{(o)}TG^{(o)}} \tag{1.22}$$

Comparing formulas (1.22) and (1.21) we obtain:

$$\tilde{T} = \tilde{G}^{(o)-1} \cdot \widetilde{G^{(o)}TG^{(o)}} \cdot \tilde{G}^{(o)-1} \tag{1.23}$$

It can be shown that on mass-shell the following equality holds

$$\tilde{T} = T$$

Derive now the equation for the amplitude $\tilde{T}$. Using the definition (1.16) we get the equation for the Fourier transform of the "two-time" Green function

$$\tilde{G} = \tilde{G}^{(o)} + \tilde{G}^{(o)} V \tilde{G} \qquad (1.24)$$

In (1.24) the multiplication is understood as a three-dimensional integration over the corresponding variables x and $\vec{p}_\perp$. Comparing (1.24) and (1.21) one can see that

$$\tilde{T}\tilde{G}^{(o)} = V\tilde{G} \qquad (1.25)$$

from which the equation for scattering amplitude $\tilde{T}$ follows:

$$\tilde{T} = V + V\tilde{G}^{(o)}\tilde{T} \qquad (1.26)$$

In the frame, where total transverse momentum is zero $\vec{P}_\perp = 0$, the equation (1.26) looks as follows:

$$\tilde{T}(P;x,\vec{p}_\perp;x',\vec{p}_\perp') = V(P;x,\vec{p}_\perp;x',\vec{p}_\perp') +$$
$$+\int_0^1 \frac{dx''}{x''(1-x'')} \int d\vec{p}_\perp'' \frac{V(P;x,\vec{p}_\perp;x'',\vec{p}_\perp'')\tilde{T}(P;x'',\vec{p}_\perp'';x',\vec{p}_\perp')}{\left[\dfrac{m_1^2+\vec{p}_\perp^{\,''2}}{x''} + \dfrac{m_2^2+\vec{p}_\perp^{\,''2}}{1-x''} - P^2 - i\varepsilon\right]} \qquad (1.27)$$

In a number of investigations (see, e.g., [48 -50]) the composite systems have been described on the basis of the so called old-fashioned three-dimensional perturbation theory in the infinite momentum frame, which has been used by Weinberg [51] in the relativistic quantum field theory. The equation (1.27) is the close analogue of the equation derived in [51]. Equation (1.27) reproduces in the lowest order of perturbation theory the equation from [51] and at the same time contains the regular method of constructing the interaction kernel in the higher orders of perturbation theory. We will not discuss this point here, but recall that as in the canonical three-dimensional approach [38] there exist the two methods of constructing of the interaction kernel (by means of the "two-time" Green function and by means of the scattering amplitude on mass-shell).

We note, however, that there exist one substantial difference between the equation derived here and equation of [51]. In the light front approach the equation is written in an arbitrary Lorentz frame and "longitudinal motion" of constituents is parametrized in terms of scale invariant and Lorentz invariant variable $x = (P/2 + p)_+/P_+$. The Weinberg equation is written in the infinite momentum frame and "longitudinal motion" is parametrized in terms of the variable $x = (P/2 + p)_z/P_z$, which is not Lorentz invariant.

3. Equation for the Many-Body Bound State Wave Function

Formalism developed in the previous sections can be generalized to the case of N relativistic interacting particles. The way of this generalization can be seen if instead of the variable x, defined by the relative momentum, two variables $x^{(1)}$ and $x^{(2)}$, defined by the individual momenta of particles

$$x^{(i)} = \frac{p_+^{(i)}}{P_+}, \qquad i = 1, 2 \tag{1.28}$$

are used. The variables $x^{(i)}$ vary in the interval $0 < x^{(i)} < 1$.

Define the Fourier transform of many-body Bethe-Salpeter amplitude

(wave function)

$$\chi_{P,\alpha}([x^{(l)}]) = \langle 0 | T\left(\varphi_1(x_\mu^{(1)})\varphi_2(x_\mu^{(2)})\ldots\varphi_N(x_\mu^{(N)})\right)|P,\alpha\rangle$$

by the following equation

$$\begin{aligned}&\delta^{(4)}(P-\sum_{l=1}^{N}p^{(l)})\,\chi_{P,\alpha}([p^{(l)}]) = \\ &= \int \prod_{l=1}^{N} d^4x^{(l)} \exp\Big[i\sum_{l=1}^{N}p^{(l)}x^{(l)}\Big]\chi_{P,\alpha}([x_\mu^{(l)}])\end{aligned} \tag{1.29}$$

where

$$[p^{(l)}] = p^{(1)},\ldots,p^{(N)}; \qquad [x_\mu^{(l)}] = x_\mu^{(1)},\ldots,x_\mu^{(N)}.$$

Here we have ascribed the Lorentz index μ to the 4-coordinats $x^{(l)}$ in order to distiguish them from the scale-invariant variables $x^{(l)}$, which will be introduced later.

Introduce the light front variables

$$P_\pm = P_0 \pm P_3; \quad p_\pm^{(l)} = p_0^{(l)} \pm p_3^{(l)}; \quad x_\pm^{(l)} = \frac{x_0^{(l)} \pm x_3^{(l)}}{2} \tag{1.30}$$

and integrate (1.29) over $\prod_{l=1}^{N} d p_-^{(l)}$. We obtain

$$\begin{aligned}&2\,\delta(P_+ - \sum_{l=1}^{N}p_+^{(l)})\,\delta^{(2)}(\vec{P}_\perp - \sum_{l=1}^{N}\vec{p}_\perp^{\,(l)})\,\Psi_{P,\alpha}([p_+^{(l)},\vec{p}_\perp^{\,(l)}]) = \\ &= (2\pi)^N \int \prod_{l=1}^{N} d^4x^{(l)}\,\delta(x_+^{(l)})\,\exp\Big[i\sum_{l=1}^{N}(p_+^{(l)}x_-^{(l)} - \vec{p}_\perp^{\,(l)}\vec{x}_\perp^{\,(l)})\Big]\chi_{P,\alpha}([x_\mu^{(l)}])\end{aligned} \tag{1.31}$$

The function $\Psi_{P,\alpha}([p_+^{(l)},\vec{p}_\perp^{\,(l)}])$ is related to the Bethe-Salpeter amplitude in the following way

$$\Psi_{P,\alpha}([p_+^{(l)},\vec{p}_\perp^{\,(l)}]) = \int_{-\infty}^{\infty}\prod_{l=1}^{N}dp_-^{(l)}\,\delta(P_- - \sum_{l=1}^{N}p_-^{(l)})\,\chi_{P,\alpha}([p^{(l)}]) \tag{1.32}$$

Introduce now the Fourier transform of the "two-time" Green function

$$\tilde{G}(P;[p_+^{(l)},\vec{p}_\perp^{(l)}];[p_+^{(l)'},\vec{p}_\perp^{(l)'}])= \tag{1.33}$$
$$=\int_{-\infty}^{\infty}\prod_{l=1}^{N}dp_-^{(l)}dp_-^{(l)'}\delta(P_- - \sum_{l=1}^{N}p_-^{(l)})\delta(P_- - \sum_{l=1}^{N}p_-^{(l)'})G(P;[p^{(l)}];[p^{(l)'}])$$

The function $G(P;[p^{(l)}];[p^{(l)'}])$ is defined by the Fourier transformation

$$G([x_\mu^{(l)}];[x_\mu^{(l)'}])=$$
$$=\langle 0|T\Big(\varphi_1(x_\mu^{(1)})\ldots\varphi_N(x_\mu^{(N)})\varphi_1^+(x_\mu^{(1)'})\ldots\varphi_N^+(x_\mu^{(N)'})\Big)|0\rangle= \tag{1.34}$$
$$=(2\pi)^{-4N}\int\prod_{l=1}^{N}d^4p^{(l)}d^4p^{(l)'}\exp\Big[-i\sum_{l=1}^{N}(p^{(l)}x^{(l)}-p^{(l)'}x^{(l)'})\Big]\times$$
$$\times G(P;[p^{(l)}];[p^{(l)'}])$$

For the case of free particles we have

$$G^{(o)}(P;[p^{(l)}];[p^{(l)'}])=\frac{i^N\prod_{l=1}^{N}\delta^{(4)}(p^{(l)}-p^{(l)'})}{\prod_{l=1}^{N}(p^{(l)2}-m^{(l)2}+i\varepsilon)} \tag{1.35}$$

Integrating both sides of (1.35) according to the definition (1.33) and ommiting the δ-function corresponding to the total 4-momentum conservation, we get [52]:

$$\tilde{G}^{(o)}(P;[p_+^{(l)},\vec{p}_\perp^{(l)}];[p_+^{(l)'},\vec{p}_\perp^{(l)'}])=$$
$$=\frac{(2i)^N(2\pi i)^{N-1}\prod_{l=1}^{N}\delta(p_+^{(l)}-p_+^{(l)'})\delta^{(2)}(\vec{p}_\perp^{(l)}-\vec{p}_\perp^{(l)'})\prod_{l=1}^{N}\Theta(x^{(l)})\Theta(1-x^{(l)})}{P_+^{N-1}\prod_{l=1}^{N}x^{(l)}\Big[P^2-\sum_{l=1}^{N}\frac{(\vec{p}_\perp^{(l)}-x^{(l)}\vec{P}_\perp)^2+m_l^2}{x^{(l)}}\Big]}= \tag{1.36}$$
$$=\tilde{G}^{(o)}(P;[p_+^{(l)},\vec{p}_\perp^{(l)}])\prod_{l=1}^{N}\delta(p_+^{(l)}-p_+^{(l)'})\delta^{(2)}(\vec{p}_\perp^{(l)}-\vec{p}_\perp^{(l)'})$$

The variables $x^{(i)}$ are defined in the following way

$$x^{(i)} = \frac{p_+^{(i)}}{P_+}, \qquad i = 1, 2, \ldots, N. \tag{1.37}$$

Thus, the function $\tilde{G}^{(0)}(P;[p_+^{(i)},\vec{p}_\perp^{(i)}])$ is defined under the conditions

$$\sum_{i=1}^{N} x^{(i)} = 1; \quad 0 < x^{(i)} < 1; \quad \sum_{i=1}^{N} \vec{p}_\perp^{(i)} = \vec{P}_\perp \tag{1.38}$$

Introduce now the inverse operator $\tilde{G}^{-1}$ by means of the relation

$$\begin{gathered}\int_0^{P_+} \prod_{i=1}^{N} dp_+^{(i)''} \int \prod_{i=1}^{N} d\vec{p}_\perp^{(i)''}\, \tilde{G}^{-1}(P;[p_+^{(i)},\vec{p}_\perp^{(i)}];[p_+^{(i)''},\vec{p}_\perp^{(i)''}]) \times \\ \times \tilde{G}(P;[p_+^{(i)''},\vec{p}_\perp^{(i)''}];[p_+^{(i)'},\vec{p}_\perp^{(i)'}]) = \prod_{i=1}^{N} \delta(p_+^{(i)} - p_+^{(i)'})\, \delta^{(2)}(\vec{p}_\perp^{(i)} - \vec{p}_\perp^{(i)'})\end{gathered} \tag{1.39}$$

and define the interaction kernel V:

$$\begin{gathered}\tilde{G}^{-1}(P;[p_+^{(i)},\vec{p}_\perp^{(i)}];[p_+^{(i)'},\vec{p}_\perp^{(i)'}]) = \\ = \tilde{G}^{(0)-1}(P;[p_+^{(i)},\vec{p}_\perp^{(i)}];[p_+^{(i)'},\vec{p}_\perp^{(i)'}]) - \\ - \frac{\delta(P_+ - \sum_{i=1}^{N} p_+^{(i)})\, \delta^{(2)}(\vec{P}_\perp - \sum_{i=1}^{N} \vec{p}_\perp^{(i)})}{(2i)^N (2\pi i)^{N-1}} V(P;[p_+^{(i)},\vec{p}_\perp^{(i)}];[p_+^{(i)'},\vec{p}_\perp^{(i)'}])\end{gathered} \tag{1.40}$$

The equation for the wave function

$$\Phi_{P,\alpha}([x^{(i)},\vec{p}_\perp^{(i)}]) = P_+^{N-1} \prod_{i=1}^{N} x^{(i)}\, \Psi_{P,\alpha}([p_+^{(i)},\vec{p}_\perp^{(i)}]) \tag{1.41}$$

looks as follows

$$\left[P^2 - \sum_{i=1}^{N} \frac{(\vec{p}_\perp^{(i)} - x^{(i)}\vec{P}_\perp)^2 + m_i^2}{x^{(i)}}\right] \Phi_{P,\alpha}([x^{(i)},\vec{p}_\perp^{(i)}]) =$$

$$= \int_0^1 \prod_{l=1}^{N} \frac{dx^{(l)'}}{x^{(l)'}} \delta(1 - \sum_{l=1}^{N} x^{(l)'}) \int \prod_{l=1}^{N} d\vec{p}_\perp^{(l)'} \delta(\vec{P}_\perp - \sum_{l=1}^{N} \vec{p}_\perp^{(l)'}) \times \qquad (1.42)$$

$$\times V(P; [p_+^{(l)}, \vec{p}_\perp^{(l)}]; [p_+^{(l)'}, \vec{p}_\perp^{(l)'}]) \Phi_{P,\alpha}([x^{(l)'}, \vec{p}_\perp^{(l)'}])$$

The formalism developed can be used for the treatment of a wide class of elementary particle and nuclear physics problems (see e.g. [53 - 55]).

Another form of the light front description of particle dynamics can be found in Refs. [56 - 60].

Chapter II

RELATIVISTIC ELASTIC FORM FACTORS AND SCATTERING AMPLITUDES FOR COMPOSITE SYSTEMS

1. Formulation of the Method

The reaction of a composite systems on a weak external perturbation corresponding to the local field $A(x)$ is described in quantum field theory by the expression [61]

$$\langle P,\alpha|\frac{\delta S}{\delta A(k)}|Q,\beta\rangle\Big|_{A=0} = (2\pi)^4\delta^{(4)}(P-Q-k)\langle P,\alpha|J(0)|Q,\beta\rangle \qquad (2.1)$$

Here $J(x)$ is the local current of the system

$$J(x)=i\,\frac{\delta S}{\delta A(x)}S^{+} \qquad (2.2)$$

$|P,\alpha\rangle$ and $|Q,\beta\rangle$ are the state vectors of composite particles with momenta P and Q and the sets of additional quantum numbers α and β, respectively, normalized in a relativistically invariant manner

$$\langle P,\alpha|Q,\beta\rangle = 2P_0(2\pi)^3\delta^{(3)}(\vec{P}-\vec{Q}) \qquad (2.3)$$

Below we suggest a method of constructing relativistically covariant form factors of composite systems in terms of "equal time" three-dimensional wave function $\Phi_{P,\alpha}([x^{(l)},\vec{p}_{\perp}^{(l)}])$.

Consider first the case of two-particle system. Introduce the quantity R defined by the vacuum expectation value of the chronologically ordered product of Heisenberg field operators of the scalar particles $\varphi_l(x_l)$ and a some local current $J(x)$:

$$R(x_1,x_2;y_1,y_2)= \\ =\langle 0|T\big(\varphi_1(x_1)\varphi_2(x_2)J(0)\varphi_1^+(y_1)\varphi_2^+(y_2)\big)|0\rangle = \\ =(2\pi)^{-16}\int d^4p_1d^4p_2d^4q_1d^4q_2\,\exp\Big[-i\sum_{j=1}^{2}(p_jx_j-q_jy_j)\Big]\times \\ \times R(p_1,p_2;q_1,q_2) \tag{2.4}$$

Introducing, as above, the relative 4-coordinates and 4-momenta

$$X=\frac{x_1+x_2}{2},\quad x=x_1-x_2\;;\qquad Y=\frac{y_1+y_2}{2},\quad y=y_1-y_2\;; \\ P=p_1+p_2,\quad p=\frac{p_1-p_2}{2}\;;\qquad Q=q_1+q_2,\quad q=\frac{q_1-q_2}{2}. \tag{2.5}$$

we rewrite expression (2.4) in the form

$$R(X,x;Y,y)= \\ =(2\pi)^{-16}\int d^4Pd^4pd^4Qd^4q\,\exp\big[-i(PX-QY+px-qy)\big]\times \\ \times R(P,p;Q,q) \tag{2.6}$$

As is known [62], the R quantity can be presented in the form

$$R=G\Gamma G, \tag{2.7}$$

or in the detailed form:

$$R(X,x;Y,y)=\int d^4X'd^4x'd^4Y'd^4y'\, G(X-X';\, x,x')\times$$
$$\times\Gamma(X',x';Y',y')G(Y'-Y;\, y,y') \qquad (2.8)$$

In the momentum space we get:

$$R(P,p:Q,q)=\int d^4p'd^4q'\, G(P;p,p')\Gamma(P,p';Q,q')G(Q;q',q) \qquad (2.9)$$

Here G is the two-particle Green function of scalar fields $\varphi_i(x_i)$ and the vertex function Γ is the sum of all two-particle irreducible diagrams for 5-point Green function (2.6)(see Fig. 1).

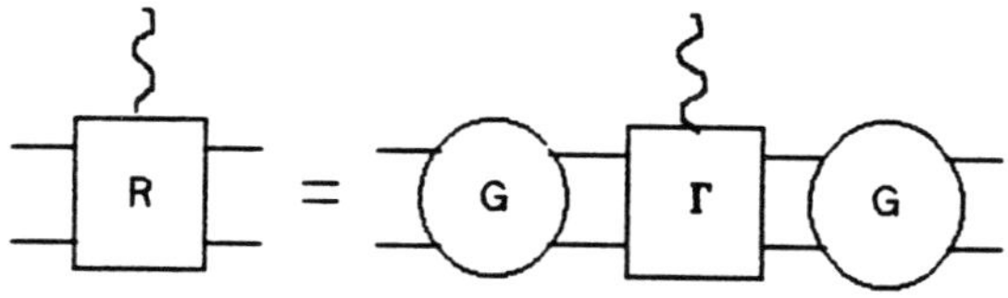

Fig. 1.

Passing to the "two-time" description in terms of light front variables, we define the quantity

$$\tilde{R}(P;p_+,\vec{p}_\perp;Q;q_+,\vec{q}_\perp)=\int_{-\infty}^{+\infty} dp_-dq_-\, R(P,p;Q,q) \qquad (2.10)$$

The quantity $\tilde{R}$ can be presented in the form

$$\tilde{R}=\tilde{G}\,\tilde{\Gamma}\,\tilde{G} \qquad (2.11)$$

or in the detailed form

$$\tilde{R}(P;p_+,\vec{p}_\perp;Q;q_+,\vec{q}_\perp)=\int_{-P_+/2}^{P_+/2}dp'_+\int d\vec{p}'_\perp\int_{-Q_+/2}^{Q_+/2}dp''_+\int d\vec{p}''_\perp\times$$
$$\times\tilde{G}(P;p_+,\vec{p}_\perp;p'_+,\vec{p}'_\perp)\tilde{\Gamma}(P;p'_+,\vec{p}'_\perp;p''_+,\vec{p}''_\perp)\tilde{G}(Q;p''_+,\vec{p}''_\perp;q_+,\vec{q}_\perp) \tag{2.12}$$

Here $\tilde{\Gamma}$ is the vertex integral operator. Let us show that the quantity $\tilde{\Gamma}$ defines the form factor of composite system. Starting from the spectral properties [63] of the 5-point Green function (2.10), it is possible to show that the quantity $\tilde{R}$ has the pole singularities near the points corresponding to the masses M_α and M_β of composite systems:

$$\tilde{R}(P;p_+,\vec{p}_\perp;Q;q_+,\vec{q}_\perp)\underset{P^2\to M_\alpha^2;\ Q^2\to M_\beta^2}{\simeq}$$
$$\simeq[i(2\pi)^4]^2\frac{\Psi_{P,\alpha}(p_+,\vec{p}_\perp)\langle P,\alpha|J(0)|Q,\beta\rangle\Psi^+_{Q,\beta}(q_+,\vec{q}_\perp)}{(P^2-M_\alpha^2)(Q^2-M_\beta^2)} \tag{2.13}$$

On the other hand, taking into account the pole singularities of the two-particle "two-time" Green function

$$\tilde{G}(P;p_+,\vec{p}_\perp;p'_+,\vec{p}'_\perp)\underset{P^2\to M_\alpha^2}{\simeq}i(2\pi)^4\frac{\Psi_{P,\alpha}(p_+,\vec{p}_\perp)\Psi^+_{P,\alpha}(p'_+,\vec{p}'_\perp)}{P^2-M_\alpha^2}, \tag{2.14}$$

we find from (2.12):

$$\tilde{R}(P;p_+,\vec{p}_\perp;Q;q_+,\vec{q}_\perp)\simeq$$
$$\simeq[i(2\pi)^4]^2\frac{\Psi_{P,\alpha}(p_+,\vec{p}_\perp)\Psi^+_{Q,\beta}(q_+,\vec{q}_\perp)}{(P^2-M_\alpha^2)(Q^2-M_\beta^2)}\int_{-P_+/2}^{P_+/2}dp_+\int d\vec{p}_\perp\int_{-Q_+/2}^{Q_+/2}dq_+\int d\vec{q}_\perp\times$$
$$\times\Psi^+_{P,\alpha}(p_+,\vec{p}_\perp)\,\tilde{\Gamma}_{\alpha\beta}(P;p_+,\vec{p}_\perp;Q;q_+,\vec{q}_\perp)\Psi_{Q,\beta}(q_+,\vec{q}_\perp) \tag{2.15}$$

where

$$\tilde{\Gamma}_{\alpha\beta}(P;p_+,\vec{p}_\perp;Q;q_+,\vec{q}_\perp)=\tilde{\Gamma}(P;p_+,\vec{p}_\perp;Q;q_+,\vec{q}_\perp)\Big|_{\substack{P^2=M_\alpha^2\\ Q^2=M_\beta^2}} \tag{2.16}$$

Comparing equations (2.13) and (2.15) we get the following expression for the matrix element of the local current J :

$$\langle P,\alpha|J(0)|Q,\beta\rangle=\int_{-P_+/2}^{P_+/2}dp_+\int d\vec{p}_\perp\int_{-Q_+/2}^{Q_+/2}dq_+\int d\vec{q}_\perp\times$$
$$\times\Psi^+_{P,\alpha}(p_+,\vec{p}_\perp)\,\tilde{\Gamma}_{\alpha\beta}(P;p_+,\vec{p}_\perp;Q;q_+,\vec{q}_\perp)\,\Psi_{Q,\beta}(q_+,\vec{q}_\perp) \tag{2.17}$$

Equations (2.15) and (2.17) give an exact expression for the vertex operator of the composite system

$$\tilde{\Gamma}_{\alpha\beta}(P;p_+,\vec{p}_\perp;Q;q_+,\vec{q}_\perp)=\lim_{\substack{P^2\to M_\alpha^2\\ Q^2\to M_\beta^2}}\int_{-P_+/2}^{P_+/2}dp'_+\int d\vec{p}\,'_\perp\int_{-Q_+/2}^{Q_+/2}dq'_+\int d\vec{q}\,'_\perp\times$$
$$\times\tilde{G}^{-1}(P;p_+,\vec{p}_\perp;p'_+,\vec{p}\,'_\perp)[\widetilde{G\Gamma G}](P;p'_+,\vec{p}\,'_\perp;Q;q'_+,\vec{q}\,'_\perp)\tilde{G}^{-1}(Q;q'_+,\vec{q}\,'_\perp;q_+,\vec{q}_\perp) \tag{2.18}$$

in terms of 4- and 5-point Green functions G and Γ. Using the perturbation theory methods for these functions one can construct the coupling constant expansion for the vertex function of composite system.

2. Elastic Form Factor in the Impulse Approximation

To demonstrate this method we consider the so-called impulse approximation for the vertex operator $\tilde{\Gamma}$ which corresponds to the limit of "weakly bound" (noninteracting) particles (see Fig. 2).

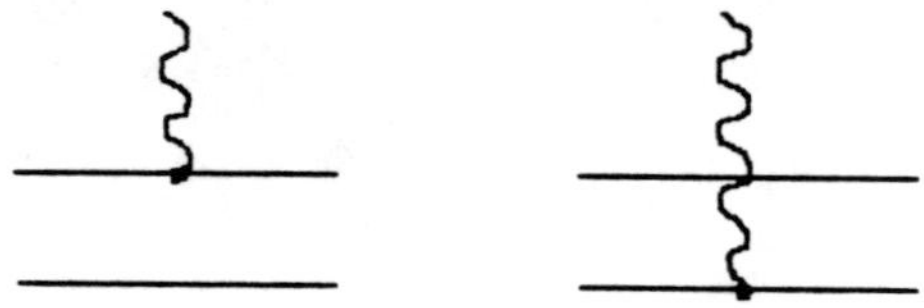

Fig. 2.

For the vertex operator corresponding to the conserved vector current we find

$$\tilde{\Gamma}_{\mu} = \tilde{\Gamma}^{(0)}_{1\mu} + \tilde{\Gamma}^{(0)}_{2\mu} \tag{2.19}$$

$$\tilde{\Gamma}^{(0)}_{1\mu} = [\tilde{G}^{(0)}]^{-1}[\widetilde{G^{(0)}\Gamma^{(0)}_{1\mu}G^{(0)}}][\tilde{G}^{(0)}]^{-1} \tag{2.20}$$

where

$$\Gamma^{(0)}_{1\mu} = (2\pi)^4 z_i(p_i+q_i)_{\mu}\delta^{(4)}(p_j-q_j)[G^{(0)}_j(p_j)]^{-1}\Big|_{i\neq j} \tag{2.21}$$

Here

$$G^{(0)}(p_j) = G^{(0)}_1(p_1)G^{(0)}_2(p_2) = i^2\prod_{i=1}^{2}(p_i^2-m_i^2)^{-1} \tag{2.22}$$

$$G^{(0)}(q_j) = G^{(0)}_1(q_1)G^{(0)}_2(q_2) = i^2\prod_{i=1}^{2}(q_i^2-m_i^2)^{-1} \tag{2.23}$$

are the two-particle Green function for free particles with masses m_i and charges z_i.

Then for the invariant form factor of the composite system defined by the relation

$$\langle P,\alpha|J_{\mu}(0)|Q,\beta\rangle = (P+Q)_{\mu}F(\Delta^2), \quad \Delta = P-Q \tag{2.24}$$

in the reference frame, in which

$$P_+ = Q_+, \qquad (P-Q)^2 = \Delta^2 = -\vec{\Delta}_\perp^2 = -(\vec{P}_\perp - \vec{Q}_\perp)^2 \tag{2.25}$$

we have [46, 50]

$$F(-\vec{\Delta}_\perp^2) = \frac{z_1(2\pi)^3}{2}\int_0^1 \frac{dx}{x(1-x)}\int d\vec{p}_\perp \Phi_{\vec{P}_\perp=0}(x,\vec{p}_\perp+(1-x)\vec{\Delta}_\perp)\Phi_{\vec{P}_\perp=0}(x,\vec{p}_\perp) + \text{similar term with } z_2 \tag{2.26}$$

Note that construction of relativistic form factors of composite systems in other versions of relativistic description of bound states is considered in Refs. [53, 63 - 65].

3. Relativistic Form Factor for the Many-Body System

Let's construct now a form factor for the relativistic many-body system in terms of the relativistic many-body wave functions $\Phi_P([x^{(l)}, \vec{p}_\perp^{(l)}])$.

Consider, as in the case of two constituents, the quantity R, which is defined by the vacuum expectation value of the chronologically ordered product of the Heisenberg field operators $\varphi_l(x_\mu^{(l)})$ and a local current $J(x)$

$$R([x_\mu^{(l)}];[y_\mu^{(l)}]) = \langle 0|T(\varphi_1(x_\mu^{(1)})\ldots\varphi_N(x_\mu^{(N)})J(0)\varphi_1^+(y_\mu^{(1)})\ldots\varphi_N^+(y_\mu^{(N)}))|0\rangle = \tag{2.27}$$

$$=(2\pi)^{-4N}\int\prod_{l=1}^N d^4p^{(l)}d^4q^{(l)}\exp[-i\sum_{l=1}^N(p^{(l)}x^{(l)}-q^{(l)}y^{(l)})]\times R([p^{(l)}];[q^{(l)}])$$

The quantity R can be presented in the form [62]

$$R = G\Gamma G \tag{2.28}$$

Multiplication in (2.28) has to be understood as an integration over the 4-coordinats of particles. G is the many-body Green function of the fields $\varphi_l(x_\mu^{(l)})$ and the vertex function Γ is defined by the sum of the irreducible diagrams of the $(2N+1)$-point function (2.27).

Proceeding now to the light front description we introduce the quantity $\tilde{R}([p_+^{(l)}, p_\perp^{(l)}];[q_+^{(l)}, q_\perp^{(l)}])$ by the relation

$$\begin{aligned} &\tilde{R}([p_+^{(l)}, p_\perp^{(l)}];[q_+^{(l)}, q_\perp^{(l)}]) = \\ &= \int_{-\infty}^{\infty} \prod_{l=1}^{N} dp_-^{(l)} dq_-^{(l)}\, \delta(P_- - \sum_{l=1}^{N} p_-^{(l)})\, \delta(Q_- - \sum_{l=1}^{N} q_-^{(l)})\, R([p^{(l)}];[q^{(l)}]) \end{aligned} \tag{2.29}$$

and write it in the form

$$\tilde{R} = \tilde{G}\tilde{\Gamma}\tilde{G} \tag{2.30}$$

Multiplication in Eq. (2.30) has to be understood in the operator sense:

$$\begin{aligned} \tilde{A}\tilde{B} = &\int_0^{Q_+} \prod_{l=1}^{N} dq_+^{(l)'}\, \delta(Q_+ - \sum_{l=1}^{N} q_+^{(l)'}) \int \prod_{l=1}^{N} d\vec{q}_\perp^{(l)'}\, \delta(\vec{Q}_\perp - \sum_{l=1}^{N} \vec{q}_\perp^{(l)'}) \times \\ &\times \tilde{A}([p_+^{(l)}, \vec{p}_\perp^{(l)}];[q_+^{(l)'}, \vec{q}_\perp^{(l)'}])\, \tilde{B}([q_+^{(l)'}, \vec{q}_\perp^{(l)'}];[q_+^{(l)}, \vec{q}_\perp^{(l)}]) \end{aligned} \tag{2.31}$$

From the spectral properties of the function $\tilde{G}$ [63] it follows that $\tilde{R}$ possesses the double pole singularities

$$\begin{aligned} &\tilde{R}([p_+^{(l)}, \vec{p}_\perp^{(l)}];[q_+^{(l)}, \vec{q}_\perp^{(l)}]) \simeq \\ &\simeq [i(2\pi)^4]^2 \frac{\Psi_{P,\alpha}([p_+^{(l)}, \vec{p}_\perp^{(l)}]) \langle P,\alpha | J(0) | Q,\beta \rangle \Psi^+_{Q,\beta}([q_+^{(l)}, \vec{q}_\perp^{(l)}])}{(P^2 - M_\alpha^2)(Q^2 - M_\beta^2)} \end{aligned} \tag{2.32}$$

in the vicinity of the points, where N-particle system forms the bound states with masses M_α and M_β and a set of other quantum numbers α and β, respectively.

On the other hand, knowing the pole singularities of the Green function

$$\tilde{G}([p_+^{(i)},\vec{p}_\perp^{(i)}];[q_+^{(i)},\vec{q}_\perp^{(i)}]) \underset{P^2\to M_\alpha^2}{\simeq} \simeq i(2\pi)^4 \frac{\Psi_{P,\alpha}([p_+^{(i)},\vec{p}_\perp^{(i)}])\Psi^+_{P,\alpha}([q_+^{(i)},\vec{q}_\perp^{(i)}])}{P^2-M_\alpha^2} \tag{2.33}$$

one can reduce the Eq. (2.32) to the form

$$\begin{aligned}
&\tilde{R}([p_+^{(i)},\vec{p}_\perp^{(i)}];[q_+^{(i)},\vec{q}_\perp^{(i)}]) \simeq \\
&\simeq [i(2\pi)^4]^2 \frac{\Psi_{P,\alpha}([p_+^{(i)},\vec{p}_\perp^{(i)}])\Psi^+_{Q,\beta}([q_+^{(i)},\vec{q}_\perp^{(i)}])}{(P^2-M_\alpha^2)(Q^2-M_\beta^2)} \times \\
&\times \int_0^{P_+N} \prod_{i=1} dp_+^{(i)} \delta(P_+-\sum_{i=1}^N p_+^{(i)}) \int \prod_{i=1}^N d\vec{p}_\perp^{(i)} \delta^{(2)}(\vec{P}_\perp-\sum_{i=1}^N \vec{p}_\perp^{(i)}) \times \\
&\times \int_0^{Q_+N} \prod_{i=1} dq_+^{(i)} \delta(Q_+-\sum_{i=1}^N q_+^{(i)}) \int \prod_{i=1}^N d\vec{q}_\perp^{(i)} \delta^{(2)}(\vec{Q}_\perp-\sum_{i=1}^N \vec{q}_\perp^{(i)}) \times \\
&\times \Psi^+_{P,\alpha}([p_+^{(i)},\vec{p}_\perp^{(i)}]) \tilde{\Gamma}_{\alpha\beta}([p_+^{(i)},\vec{p}_\perp^{(i)}];[q_+^{(i)},\vec{q}_\perp^{(i)}]) \Psi_{Q,\beta}([q_+^{(i)},\vec{q}_\perp^{(i)}])
\end{aligned} \tag{2.34}$$

Here

$$\begin{aligned}
&\tilde{\Gamma}_{\alpha\beta}([p_+^{(i)},\vec{p}_\perp^{(i)}];[q_+^{(i)},\vec{q}_\perp^{(i)}]) = \\
&= \lim_{P^2\to M_\alpha^2,\, Q^2\to M_\beta^2} \int_0^{P_+N} \prod_{i=1} dp_+^{(i)'} \delta(P_+-\sum_{i=1}^N p_+^{(i)'}) \int \prod_{i=1}^N d\vec{p}_\perp^{(i)'} \delta^{(2)}(\vec{P}_\perp-\sum_{i=1}^N \vec{p}_\perp^{(i)'}) \times \\
&\times \int_0^{Q_+N} \prod_{i=1} dq_+^{(i)'} \delta(Q_+-\sum_{i=1}^N q_+^{(i)'}) \int \prod_{i=1}^N dq_\perp^{(i)'} \delta^{(2)}(\vec{Q}_\perp-\sum_{i=1}^N \vec{q}_\perp^{(i)'}) \times \\
&\times \tilde{G}^{-1}([p_+^{(i)},\vec{p}_\perp^{(i)}];[p_+^{(i)'},\vec{p}_\perp^{(i)'}]) \times \\
&\times [\widetilde{G\Gamma G}]([p_+^{(i)'},\vec{p}_\perp^{(i)'}];[q_+^{(i)'},\vec{q}_\perp^{(i)'}]) \tilde{G}^{-1}([q_+^{(i)'},\vec{q}_\perp^{(i)'}];[q_+^{(i)},q_\perp^{(i)}])
\end{aligned} \tag{2.35}$$

Comparing (2.32) with the (2.34) we get the following expression for the matrix element of the bound state current:

$$\langle P,\alpha|J(0)|Q,\beta\rangle =$$

$$= \int_0^{P_+} \prod_{i=1}^{N} dp_+^{(i)} \delta(P_+ - \sum_{i=1}^{N} p_+^{(i)}) \int \prod_{i=1}^{N} d\vec{p}_\perp^{(i)} \delta^{(2)}(\vec{P}_\perp - \sum_{i=1}^{N} \vec{p}_\perp^{(i)}) \times$$

$$\times \int_0^{Q_+} \prod_{i=1}^{N} dq_+^{(i)} \delta(Q_+ - \sum_{i=1}^{N} q_+^{(i)}) \int \prod_{i=1}^{N} d\vec{q}_\perp^{(i)} \delta^{(2)}(\vec{Q}_\perp - \sum_{i=1}^{N} \vec{q}_\perp^{(i)}) \times \tag{2.36}$$

$$\times \Psi^+_{P,\alpha}([p_+^{(i)}, \vec{p}_\perp^{(i)}]) \tilde{\Gamma}_{\alpha\beta}([p_+^{(i)}, \vec{p}_\perp^{(i)}]; [q_+^{(i)}, \vec{q}_\perp^{(i)}]) \Psi_{Q,\beta}([q_+^{(i)}, \vec{q}_\perp^{(i)}])$$

The vertex operator $\tilde{\Gamma}_{\alpha\beta}$ can be constructed, using, for instance, perturbation theory methods of quantum field theory. Phenomenological vertex operators can also be used. Here we consider the so-called "impulse approximation". In this case

$$\tilde{\Gamma}_\mu = \sum_{i=1}^{N} (\tilde{\Gamma}_\mu^{(o)})_i \tag{2.37}$$

$$(\tilde{\Gamma}_\mu^{(o)})_i = [\tilde{G}^{(o)}]^{-1} [\widehat{G^{(o)}(\Gamma_\mu^{(o)})_i G^{(o)}}] [\tilde{G}^{(o)}]^{-1} \tag{2.38}$$

Here

$$(\Gamma_\mu^{(o)})_i = (2\pi)^4 z_i (p^{(i)} + q^{(i)})_\mu \prod_{\substack{j=1 \\ i \neq j}}^{N} \delta(p^{(j)} - q^{(j)}) [G^{(o)}(q^{(j)})]^{-1} \tag{2.39}$$

$$G^{(o)}([p^{(i)}]) = i^N \prod_{i=1}^{N} (p^{(i)2} - m_i^2 + i\varepsilon)^{-1} \tag{2.40}$$

and z_i is the charge of i-th particle.

Performing the integration in the Eq. (2.38) according to the definition (2.29), passing to the reference frame, where

$$P_+ = Q_+, \quad \vec{Q}_\perp = 0, \quad \Delta^2 = (P-Q)^2 = -\vec{\Delta}_\perp^2 = -(\vec{P}_\perp - \vec{Q}_\perp)^2 \tag{2.41}$$

and using the transformation properties of wave functions

$$\Phi_{P,\alpha}([x^{(l)},\vec{p}_{\perp}^{(l)}])=\Phi_{\vec{P}_{\perp}=0,\alpha}([x^{(l)},\vec{p}_{\perp}^{(l)}-x^{(l)}\vec{P}_{\perp}]) \tag{2.42}$$

we obtain [52]:

$$\langle P,\alpha|J_{\mu}(0)|Q,\beta\rangle=\sum_{k=1}^{N}\langle P,\alpha|J_{\mu}(0)|Q,\beta\rangle_{k} \tag{2.43}$$

where, for instance

$$\begin{aligned}
&\langle P,\alpha|J_{+}(0)|Q,\beta\rangle_{k}=(P_{+}+Q_{+})F_{k}(-\vec{\Delta}_{\perp}^{2})=\\
&=\frac{-(2\pi)^{4}z_{k}(P_{+}+Q_{+})}{(2i)^{N+1}(2\pi i)^{N-1}}\int_{0}^{1}\prod_{l=1}^{N}\frac{dx^{(l)}}{x^{(l)}}\,\delta(1-\sum_{l=1}^{N}x^{(l)})\int\prod_{l=1}^{N}d\vec{p}_{\perp}^{(l)}\,\delta^{(2)}(\sum_{l=1}^{N}\vec{p}_{\perp}^{(l)})\times\\
&\times\Phi^{+}_{\vec{P}_{\perp}=0,\alpha}([x^{(l)},\vec{p}_{\perp}^{(l)}-x^{(l)}\vec{\Delta}_{\perp}]_{l\neq k},x^{(k)},\vec{p}_{\perp}^{(k)}+(1-x^{(k)})\vec{\Delta}_{\perp})\times\\
&\times\Phi_{\vec{P}_{\perp}=0,\beta}([x^{(l)},\vec{p}_{\perp}^{(l)}])
\end{aligned} \tag{2.44}$$

The wave functions $\Phi_{\vec{P}_{\perp}=0,\alpha}([x^{(l)},\vec{p}_{\perp}^{(l)}])$ are defined by the (1.41). Taking into account the normalization condition for the wave functions

$$\begin{aligned}
&i(2\pi)^{4}\int_{0}^{P_{+}}\prod_{l=1}^{N}dp_{+}^{(l)}\delta(P_{+}-\sum_{l=1}^{N}p_{+}^{(l)})\int\prod_{l=1}^{N}d\vec{p}_{\perp}^{(l)}\delta^{(2)}(\vec{P}_{\perp}-\sum_{l=1}^{N}\vec{p}_{\perp}^{(l)})\times\\
&\times\int_{0}^{Q_{+}}\prod_{l=1}^{N}dq_{+}^{(l)}\delta(Q_{+}-\sum_{l=1}^{N}q_{+}^{(l)})\int\prod_{l=1}^{N}d\vec{q}_{\perp}^{(l)}\delta^{(2)}(\vec{Q}_{\perp}-\sum_{l=1}^{N}\vec{q}_{\perp}^{(l)})\times\\
&\times\Psi^{+}_{P,\alpha}([p_{+}^{(l)},\vec{p}_{\perp}^{(l)}])\frac{\partial\tilde{G}^{-1}(P;[p_{+}^{(l)},\vec{p}_{\perp}^{(l)}];[q_{+}^{(l)},\vec{q}_{\perp}^{(l)}])}{\partial P^{2}}\Psi_{Q,\beta}([q_{+}^{(l)},\vec{q}_{\perp}^{(l)}])=1
\end{aligned} \tag{2.45}$$

for $\vec{\Delta}_{\perp}=0$ we get:

$$F(\Delta^{2}=0)=\frac{1}{2}\sum_{l=1}^{N}z_{k} \tag{2.46}$$

Thus, the form factor at zero momentum transfer is normalized to the total charge of the system. Note, that general problems of normalization of three-dimensional relativistic wave functions have been considered in Ref. [66].

Thus Equations (2.43) and (2.44) define the form factors of a many-body system in terms of the light front relativistic wave functions $\Phi([x^{(i)}, \vec{p}_{\perp}^{\,(i)}])$.

4. Scattering of Relativistic Composite Systems

Experimental study of high energy processes during the last decades revealed a number of scaling properties of observable quantities. Many of these properties can be understood on the basis of the composite quark-parton structure of elementary particles. In particular, the asymptotic scaling property of differential cross section of hadron-hadron scattering

$$\left.\frac{d\sigma}{dt}\right|_{\substack{s \to \infty \\ |t/s| = \text{const.}}} \sim \frac{1}{s^N} f(\cos\theta_s) \tag{2.47}$$

where N is integer number, can be explained in the framework of dimensional analysis and assumption on three-quark structure of baryons and quark-antiquark structure of mesons (quark counting rules) [67, 68].

In connection with the development of composite models of elementary particles a problem of the description of their interactions becomes of special interest. Study of interactions of relativistic composite systems is important also in connection with current and future experiments with beams of relativistic nuclei. Here we outline a method for the treatment of problems of that kind [69].

Below we present a description of the scattering of two composite particles. It will be shown that some simple assumptions on the hadron interactions in the scattering process allow to reproduce the results of quark counting rules.

Consider the eight-point Green function G:

$$
\begin{aligned}
&G(x_1,x_2,x_3,x_4;y_1,y_2,y_3,y_4)=\\
&=\langle 0|T(\varphi_1(x_1)\varphi_2(x_2)\varphi_3(x_3)\varphi_4(x_4)\varphi_1^+(y_1)\varphi_2^+(y_2)\varphi_3^+(y_3)\varphi_4^+(y_4))|0\rangle=\\
&=[(2\pi)^4]^{-8}\int\prod_{i=1}^{4}d^4p_i d^4q_i \exp\Big[-i\sum_{i=1}^{4}(p_i x_i-q_i y_i)\Big]G(p_1,p_2,p_3,p_4;q_1,q_2,q_3,q_4)=\\
&=[(2\pi)^4]^{-8}\int d^4P^{(12)}d^4p^{(12)}d^4P^{(34)}d^4p^{(34)}d^4Q^{(12)}d^4q^{(12)}d^4Q^{(34)}d^4q^{(34)}\times\\
&\times\exp\Big[-i(P^{(12)}X^{(12)}+P^{(34)}X^{(34)}+p^{(12)}x^{(12)}+p^{(34)}x^{(34)}-\\
&-Q^{(12)}Y^{(12)}-Q^{(34)}Y^{(34)}-q^{(12)}y^{(12)}-q^{(34)}y^{(34)})\Big]\times\\
&\times G(P^{(12)},p^{(12)};P^{(34)},p^{(34)};Q^{(12)},q^{(12)};Q^{(34)},q^{(34)})
\end{aligned}
\tag{2.48}
$$

In (2.48) the momenta $P^{(12)}$, $p^{(12)}, P^{(34)}, p^{(34)}, Q^{(12)}, q^{(12)}, Q^{(34)}, q^{(34)}$ are introduced according to the following Equations

$$
\begin{aligned}
&P^{(12)}=p_1+p_2,\quad p^{(12)}=\frac{p_1-p_2}{2},\quad P^{(34)}=p_3+p_4,\quad p^{(34)}=\frac{p_3-p_4}{2},\\
&Q^{(12)}=q_1+q_2,\quad q^{(12)}=\frac{q_1-q_2}{2},\quad Q^{(34)}=q_3+q_4,\quad q^{(34)}=\frac{q_3-q_4}{2}.
\end{aligned}
\tag{2.49}
$$

Pasing now to the "two-time" description, we introduce the light front variables and define the quantity

$$
\begin{aligned}
&\tilde G(P^{(12)},p_+^{(12)},\vec p_\perp^{(12)};P^{(34)},p_+^{(34)},\vec p_\perp^{(34)};Q^{(12)},q_+^{(12)},\vec q_\perp^{(12)};Q^{(34)},q_+^{(34)},\vec q_\perp^{(34)})=\\
&=\int_{-\infty}^{\infty}dp_-^{(12)}dp_-^{(34)}dq_-^{(12)}dq_-^{(34)}\,G(P^{(12)},p^{(12)};P^{(34)},p^{(34)};Q^{(12)},q^{(12)};Q^{(34)},q^{(34)})
\end{aligned}
\tag{2.50}
$$

where $p_{\pm}=p_0 \pm p_3$, $\vec{p}_{\perp}=(p_1, p_2)$.

Introduce now the quantity M by the Equation

$$
\begin{gathered}
G(P^{(12)},p^{(12)};P^{(34)},p^{(34)};Q^{(12)},q^{(12)};Q^{(34)},q^{(34)})= \\
=\int d^4p^{(12)'}d^4p^{(34)'}d^4q^{(12)'}d^4q^{(34)'}\times \\
\times G_{12}(P^{(12)},p^{(12)},p^{(12)'})G_{34}(P^{(34)},p^{(34)},p^{(34)'})\times \\
\times M(P^{(12)},p^{(12)'};P^{(34)},p^{(34)'};Q^{(12)},q^{(12)'};Q^{(34)},q^{(34)'})\times \\
\times G_{12}(Q^{(12)},q^{(12)'},q^{(12)})G_{34}(Q^{(34)},q^{(34)'},q^{(34)})\equiv(G_{12}G_{34})M(G_{12}G_{34})
\end{gathered}
\tag{2.51}
$$

The quantity $\tilde{G}$ can be presented in the form (here and in what follows we omit the arguments which are related to the relative momenta and this will not cause any misunderstanding):

$$
\begin{gathered}
\tilde{G}(P^{(12)},P^{(34)},Q^{(12)},Q^{(34)})= \\
=\tilde{G}_{12}(P^{(12)})\tilde{G}_{34}(P^{(34)})*\tilde{M}(P^{(12)},P^{(34)},Q^{(12)},Q^{(34)})*\tilde{G}_{12}(Q^{(12)})\tilde{G}_{34}(Q^{(34)})
\end{gathered}
\tag{2.52}
$$

The symbol * in the (2.52) has to be understood in the following sense

$$
\begin{gathered}
\tilde{A}*\tilde{B}=\int_{-P_+^{(12)}/2}^{P_+^{(12)}/2} dp_+^{(12)}\int_{-P_+^{(34)}/2}^{P_+^{(34)}/2} dp_+^{(34)}\int d\vec{p}_{\perp}^{(12)}\int d\vec{p}_{\perp}^{(34)}\times \\
\times\tilde{A}(\ldots,p_+^{(12)},\vec{p}_{\perp}^{(12)};p_+^{(34)},\vec{p}_{\perp}^{(34)})\tilde{B}(p_+^{(12)},\vec{p}_{\perp}^{(12)};p_+^{(34)},\vec{p}_{\perp}^{(34)},\ldots)
\end{gathered}
\tag{2.53}
$$

and dots correspond to the set of other arguments on which the operators $\tilde{A}$ and $\tilde{B}$ can depend.

Knowing the pole singularities of the two-particle Green functions $\tilde{G}_{12}$, $\tilde{G}_{34}$ one can show that in the vicinity of these poles the function $\tilde{G}$ looks as follows:

$$\tilde{G}(P^{(12)},P^{(34)},Q^{(12)},Q^{(34)})\simeq$$

$$\simeq[i(2\pi)^4]^4\frac{\Psi_{12}(P^{(12)})\Psi_{34}(P^{(34)})\Psi^+_{12}(Q^{(12)})\Psi^+_{34}(Q^{(34)})}{\left(P^{(12)2}-M^2_{12}\right)\left(P^{(34)2}-M^2_{34}\right)\left(Q^{(12)2}-M'^2_{12}\right)\left(Q^{(34)2}-M'^2_{34}\right)}\times \tag{2.54}$$

$$\times\Psi^+_{12}(P^{(12)})\Psi^+_{34}(P^{(34)})*\tilde{M}(P^{(12)},P^{(34)},Q^{(12)},Q^{(34)})*\Psi_{12}(Q^{(12)})\Psi_{34}(Q^{(34)})$$

Here

$$\tilde{M}_{1234}=\lim_{\substack{P^{(12)2}\to M^2_{12},\,P^{(34)2}\to M^2_{34}\\ Q^{(12)2}\to M'^2_{12},\,Q^{(34)2}\to M'^2_{34}}}\tilde{G}^{-1}_{12}(P^{(12)})\tilde{G}^{-1}_{34}(P^{(34)})* \tag{2.55}$$

$$*\widetilde{G_{12}G_{34}MG_{12}G_{34}}(P^{(12)},P^{(34)},Q^{(12)},Q^{(34)})*\tilde{G}^{-1}_{12}(Q^{(12)})\tilde{G}^{-1}_{34}(Q^{(34)})$$

$M^2_{12},M^2_{34},M'^2_{12},M'^2_{34}$ are the masses of corresponding states.

From Equations (2.52) and (2.54) we get the following expression for the scattering amplitude

$$T(P^{(12)},P^{(34)},Q^{(12)},Q^{(34)})=$$

$$=\Psi^+_{12}(P^{(12)})\Psi^+_{34}(P^{(34)})*\tilde{M}(P^{(12)},P^{(34)},Q^{(12)},Q^{(34)})*\Psi_{12}(Q^{(12)})\Psi_{34}(Q^{(34)}) \tag{2.56}$$

The Eq. (2.56) gives a general expression for the scattering amplitude in the case of scattering of composite particles. The detailed form of the scattering amplitude depends on the interaction mechanism in the intermediate state and on a specific form of the wave functions of the scattered objects.

5. Gluon Exchange in the Intermediate State

Below we consider two possible mechanisms: 1). scattering via the exchange of some intermediate particle (Fig. 3), 2). constituent interchange

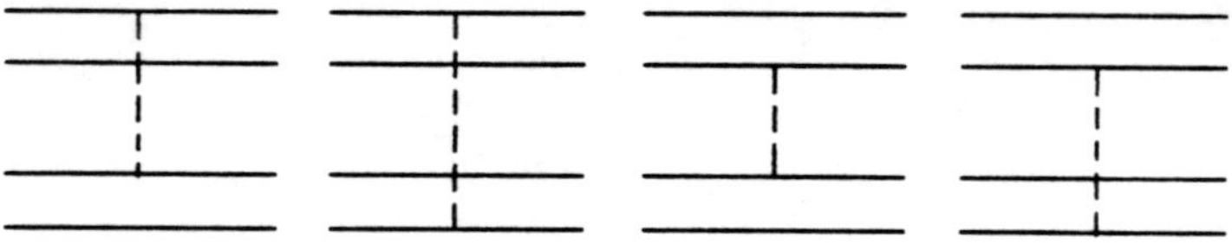

Fig. 3.

mechanism (Fig. 4). For the sake of simplicity we consider the case of scalar particles. It can be shown that in the first case in the reference frame, where

$$P_+^{(12)} = P_+^{(34)}, \quad Q_+^{(12)} = Q_+^{(34)}, \quad t = (P^{(12)} - Q^{(12)})^2 = -\vec{\Delta}_\perp^{(t)2}$$

one has the following result for the scattering amplitude

$$T = \frac{g^2}{4(\vec{\Delta}_\perp^{(t)2} + \mu^2)} [F_{12}^{(1)}(t) F_{34}^{(1)}(t) + F_{12}^{(1)}(t) F_{34}^{(2)}(t) + F_{12}^{(2)}(t) F_{34}^{(1)}(t) + F_{12}^{(2)}(t) F_{34}^{(2)}(t)] \tag{2.57}$$

where $F_{12}^{(i)}$, $F_{34}^{(i)}$ are scalar form factors of the scattered particles, which are expressed in terms of the light front wave functions in the following way:

$$\begin{aligned}
F_{12}^{(1)}(t) &= \frac{(2\pi)^3}{2} \int_0^1 \frac{dx}{x^2(1-x)} \int d\vec{p}_\perp \Phi^+_{\vec{P}_\perp^{(12)}=0}(x, \vec{p}_\perp + (1-x)\vec{\Delta}_\perp^{(t)}) \Phi_{\vec{P}_\perp^{(12)}=0}(x, \vec{p}_\perp) \\
F_{34}^{(1)}(t) &= \frac{(2\pi)^3}{2} \int_0^1 \frac{dx}{x^2(1-x)} \int d\vec{p}_\perp \Phi^+_{\vec{P}_\perp^{(34)}=0}(x, \vec{p}_\perp - (1-x)\vec{\Delta}_\perp^{(t)}) \Phi_{\vec{P}_\perp^{(34)}=0}(x, \vec{p}_\perp) \\
F_{12}^{(2)}(t) &= \frac{(2\pi)^3}{2} \int_0^1 \frac{dx}{x(1-x)^2} \int d\vec{p}_\perp \Phi^+_{\vec{P}_\perp^{(12)}=0}(x, \vec{p}_\perp - x\vec{\Delta}_\perp^{(t)}) \Phi_{\vec{P}_\perp^{(12)}=0}(x, \vec{p}_\perp) \\
F_{34}^{(2)}(t) &= \frac{(2\pi)^3}{2} \int_0^1 \frac{dx}{x(1-x)^2} \int d\vec{p}_\perp \Phi^+_{\vec{P}_\perp^{(34)}=0}(x, \vec{p}_\perp + x\vec{\Delta}_\perp^{(t)}) \Phi_{\vec{P}_\perp^{(34)}=0}(x, \vec{p}_\perp)
\end{aligned} \tag{2.58}$$

6. Constituent Interchange Mechanism

Considering the constituent interchange mechanism [50] (Fig. 4) one gets the following expression for the scattering amplitude

$$
\begin{aligned}
T = \frac{-1}{2(2\pi)^3} \int_0^1 \frac{dx}{x^2(1-x)^2} \int d\vec{p}_\perp \, \Phi^{+(12)}_{\vec{P}_\perp=0}(x, \vec{p}_\perp - x\vec{\Delta}^{(u)}_\perp + (1-x)\vec{\Delta}^{(t)}_\perp) \, \Phi^{+(34)}_{\vec{P}_\perp=0}(x, \vec{p}_\perp) \times \\
\times [M^2_{12} + M^2_{34} - S(x, \vec{p}_\perp + x\vec{\Delta}^{(t)}_\perp - (1-x)\vec{\Delta}^{(u)}_\perp) - S(x, \vec{p}_\perp)] \times \qquad (2.59) \\
\times \Phi^{(12)}_{\vec{P}_\perp=0}(x, \vec{p}_\perp - x\vec{\Delta}^{(u)}_\perp) \, \Phi^{(34)}_{\vec{P}_\perp=0}(x, \vec{p}_\perp + (1-x)\vec{\Delta}^{(t)}_\perp)
\end{aligned}
$$

where $\vec{\Delta}^{(t)}_\perp = -t$, $\vec{\Delta}^{(u)}_\perp = -u$.

a b

Fig. 4.

Here the notation has been introduced

$$
S(x, \vec{p}_\perp) = \frac{m_1^2 + \vec{p}_\perp^{\,2}}{1-x} + \frac{m_2^2 + \vec{p}_\perp^{\,2}}{x} \qquad (2.60)
$$

and the following properties of wave functions

$$
\begin{aligned}
\Phi(x, \vec{p}_\perp) &= \Phi(x, -\vec{p}_\perp) \\
\Phi(x, \vec{p}_\perp) &= \Phi(1-x, \vec{p}_\perp)
\end{aligned} \qquad (2.61)
$$

have been used.

Let us choose now the wave functions of the composite particles in the form

$$\Phi_N(x,\vec{p}_\perp)=\frac{\varphi_N(x)}{[S(x,\vec{p}_\perp)]^N}, \quad N=A,B,C,D \tag{2.62}$$

A, B and C, D denote the hadrons before and after the scattering and corresponding powers, respectively.

Inserting the wave functions (2.62) into the Eq. (2.59) for the scattering amplitude one gets in the asymptotic region

$$T \underset{\substack{s\to\infty\\ |t|\to\infty}}{\sim} \frac{1}{s^{A+C+D-1}}\left(\frac{1+z}{2}\right)^{-C}\left(\frac{1-z}{2}\right)^{-D} f(z) \tag{2.63}$$

where

$$f(z)=\int_0^1 \frac{dx\, \varphi_A^+(x)\tilde{\varphi}_B^+(x)\varphi_C(x)\varphi_D(x)}{\left[(1-x)^2\frac{1-z}{2}+x^2\frac{1+z}{2}\right]^A}\left[(1-x)^2\frac{1+z}{2}+x^2\frac{1-z}{2}\right]\times$$
$$\times[x(1-x)]^{A+B+C+D-3}\, x^{-2C}(1-x)^{-2D}, \tag{2.64}$$

$$\tilde{\varphi}_B^+(x)=\frac{-1}{(2\pi)^3}\int d\vec{p}_\perp\, \Phi_B^+(x,\vec{p}_\perp)[x(1-x)]^{-B}$$

$z=\cos\vartheta_s$, where ϑ_s is the scattering angle in the c.m.s.

$$-t\simeq\frac{s}{2}(1-z), \qquad -u\simeq\frac{s}{2}(1+z)$$

Equation (2.63) is in close connection with the results of quark counting rules [67,68].

The contribution to the scattering amplitude from the annihilation diagrams of the type given in Fig. 4b in the lowest order of perturbation theory is identically equal to zero due to the projective properties of the wave functions. Instead of calculating the higher orders one can use the $s\to u$ crossing symmetry property of the physical scattering amplitude.

The straightforward calculation of the diagram of Fig. 4b, assuming the following property for the $\tilde{M}$ operator

$$\tilde{M}(P^{(12)},x^{(12)},\vec{p}_{\perp}^{(12)};P^{(34)},x^{(34)},\vec{p}_{\perp}^{(34)};Q^{(12)},y^{(12)},\vec{q}_{\perp}^{(12)};Q^{(34)},y^{(34)},\vec{q}_{\perp}^{(34)})=$$

$$=\tilde{M}(P^{(12)},-x^{(12)},\vec{p}_{\perp}^{(12)};P^{(34)},-x^{(34)},\vec{p}_{\perp}^{(34)};Q^{(12)},-y^{(12)},\vec{q}_{\perp}^{(12)};Q^{(34)},-y^{(34)},\vec{q}_{\perp}^{(34)})$$

when $-1<x^{(12)}<0$, $-1<x^{(34)}<0$, $-1<y^{(12)}<0$, $-1<y^{(34)}<0$,

leads to the following expression for scattering amplitude:

$$T \underset{\substack{s\to\infty \\ |t|\to\infty}}{\sim} \frac{1}{s^{A+C+D-1}}(1-z)^{-D}\tilde{f}(z) \tag{2.65}$$

In this case the function $\tilde{f}(z)$ in (2.65) is of the form:

$$\tilde{f}(z)=2^{-A-C+D+1}\int_0^1 dx\,\frac{\varphi_A^+(x)\tilde{\varphi}_B^+(x)\varphi_C(x)\varphi_D(x)}{[4x^2+(1-x)^2(1-z)]^{C-1}}\times$$
$$\times\,[x(1-x)]^{A+B+C+D-3}x^{-2A}(1-x)^{-2D} \tag{2.66}$$

It is interesting to investigate the dependence of the power of s in Eqs. (2.63) and (2.65) on the number and on spins of the constituents, to study in more detail the wave functions of the scattered objects and to compare the corresponding results with experimental data. Note, that angular dependence of differential cross sections of elastic processes in the framework of the quark counting rules has been studied in Ref. [70]. Derivation of quark counting rules for inclusive processes can be found in Refs. [71, 72].

Chapter III

DEEP INELASTIC FORM FACTORS OF COMPOSITE SYSTEMS AND MULTIQUARK STATES IN NUCLEI

The great interest to deep inelastic interaction processes is caused by the possibility of studying the internal structure of hadrons and nuclei experimentally and checking different theoretical models based on the assumptions about composite nature of strongly interacting particles. The main part of experimentally observed properties of these processes (in particular, the scale properties of structure functions) have been explained in the framework of composite quark-parton models of hadrons, in which the hadron is considered as bound state of some paralelly moving pointlike constituents [73 - 76]. Interaction between constituents and their transverse motion inside hadron is neglected.

More precise measurements in wider range of kinematic variables have led to the discovery of deviations from exact scale invariance in the behaviour of structure functions [77 - 81]. There were attempts to explain these deviations on the kinematical (search for new scale-invariant variables [82]) and dynamical (taking into account higher chromodynamical corrections [83 - 85]) basis.

Here we incorporate the transverse motion of constituents in the composite system, which leads to the violation of Bjorken scaling of structure functions. In this approach, hadrons are considered as bound states of quarks, described in terms of light front wave functions.

1. Construction of Tensor $W_{\mu\nu}$.

Consider the quantity $R_{\mu\nu}$, which is defined by the vacuum expectation value of the chronological T_+-ordered product of the Heisenberg field operators $\varphi_i(x_\mu^{(i)})$ and local currents J_μ and J_ν:

$$
\begin{aligned}
&R_{\mu\nu}([x_\mu^{(i)}];[y_\mu^{(i)}];z)= \\
&=\langle 0|T\left(\varphi_1(x_\mu^{(1)})\ldots\varphi_N(x_\mu^{(N)})J_\mu(z)J_\nu(0)\varphi_1^+(y_\mu^{(1)})\ldots\varphi_N^+(y_\mu^{(N)})\right)|0\rangle= \\
&=(2\pi)^{-8N}\int\prod_{i=1}^{N}d^4p^{(i)}d^4q^{(i)}\exp\left[-i\sum_{i=1}^{N}(p^{(i)}x^{(i)}-q^{(i)}y^{(i)})\right]\times \\
&\times R_{\mu\nu}([p^{(i)}];[q^{(i)}];z)
\end{aligned}
\tag{3.1}
$$

Here $[x_\mu^{(i)}]$, $[y_\mu^{(i)}]$, $[p^{(i)}]$, $[q^{(i)}]$ are the sets of corresponding 4-vectors.

The quantity $R_{\mu\nu}$ can be presented as (see Fig. 5)

$$R_{\mu\nu}=G\,\Gamma_{\mu\nu}\,G$$

where G is the N-particle Green function of fields $\varphi_i(x_\mu^{(i)})$:

$$
\begin{aligned}
&G([x_\mu^{(i)}];[y_\mu^{(i)}])= \\
&=\langle 0|T\left(\varphi_1(x_\mu^{(1)})\ldots\varphi_N(x_\mu^{(N)})\,\varphi_1^+(y_\mu^{(1)})\ldots\varphi_N^+(y_\mu^{(N)})\right)|0\rangle
\end{aligned}
\tag{3.2}
$$

and "two-photon" vertex function $\Gamma_{\mu\nu}$ is defined by the sum of the irreducible diagrams with $2N+2$ points (legs).

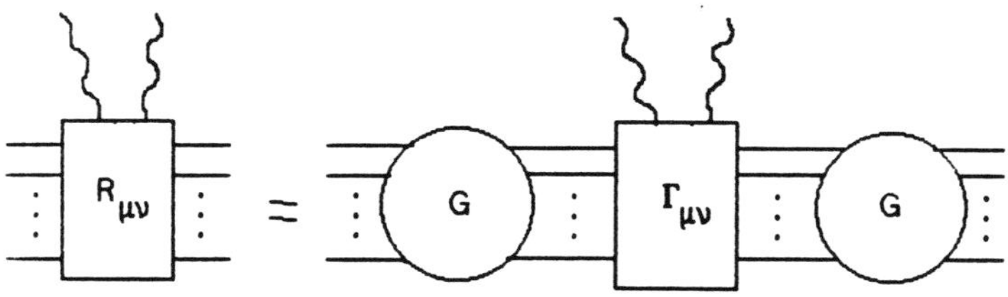

Fig. 5.

Introduce now the three-dimensional quantity $\tilde{R}_{\mu\nu}$ equating all $x_+^{(i)}=x_+$ and $y_+^{(i)}=y_+$ in (3.1)

$$\tilde{R}_{\mu\nu}(x_+,[x_-^{(i)},\vec{x}_\perp^{(i)}];y_+,[y_-^{(i)},\vec{y}_\perp^{(i)}];z)=$$

$$=\langle 0|T\big(\varphi_1(x_+,x_-^{(1)},\vec{x}_\perp^{(1)})\ldots\varphi_N(x_+,x_-^{(N)},\vec{x}_\perp^{(N)})J_\mu(z)J_\nu(0)\times$$

$$\times\varphi_1^+(y_+,y_-^{(1)},\vec{y}_\perp^{(1)})\ldots\varphi_N^+(y_+,y_-^{(N)},\vec{y}_\perp^{(N)})\big)|0\rangle=$$

$$=(2\pi)^{-8N}\int\prod_{i=1}^{N}\Big(\frac{1}{2}dp_+^{(i)}d\vec{p}_\perp^{(i)}\Big)\Big(\frac{1}{2}dq_+^{(i)}d\vec{q}_\perp^{(i)}\Big)\tilde{R}_{\mu\nu}([p_+^{(i)},\vec{p}_\perp^{(i)}];[q_+^{(i)},\vec{q}_\perp^{(i)}];z)\times \tag{3.3}$$

$$\times\exp\Big[-\frac{i}{2}(x_+P_--y_+Q_-)-i\sum_{i=1}^{N}\Big(\frac{1}{2}p_+^{(i)}x_-^{(i)}-\vec{p}_\perp^{(i)}\vec{x}_\perp^{(i)}\Big)+i\sum_{i=1}^{N}\Big(\frac{1}{2}q_+^{(i)}y_-^{(i)}-\vec{q}_\perp^{(i)}\vec{y}_\perp^{(i)}\Big)\Big]$$

Fourier transforms of $R_{\mu\nu}$ and $\tilde{R}_{\mu\nu}$ are related to each other in the following way:

$$\tilde{R}_{\mu\nu}([p_+^{(i)},\vec{p}_\perp^{(i)}];[q_+^{(i)},\vec{q}_\perp^{(i)}];z)=$$

$$=\int_{-\infty}^{\infty}\prod_{i=1}^{N}dp_-^{(i)}dq_-^{(i)}\,\delta\Big(P_--\sum_{i=1}^{N}p_-^{(i)}\Big)\delta\Big(Q_--\sum_{i=1}^{N}q_-^{(i)}\Big)R_{\mu\nu}([p^{(i)}];[q^{(i)}];z) \tag{3.4}$$

Single out now the contribution of N-particle bound states in matrix element (3.3) expressing the T-product via Θ-functions and using the compliteness of physical states. Using the integral representation of Θ-function we get that the quantity $\tilde{R}_{\mu\nu}$ has the double pole singularities in the vicinity of the points corresponding to bound states with masses M_α and M_β and a set of other quantum numbers α and β, respectively:

$$\tilde{R}_{\mu\nu}([p_+^{(l)},\vec{p}_\perp^{(l)}];[q_+^{(l)},\vec{q}_\perp^{(l)}];z)\simeq$$
$$\simeq\left[\frac{i}{(2\pi)^4}\right]^2\frac{\Psi_P([p_+^{(l)},\vec{p}_\perp^{(l)}])\langle P,\alpha|T(J_\mu(z)J_\nu(0))|Q,\beta\rangle\Psi_Q^+([q_+^{(l)},\vec{q}_\perp^{(l)}])}{(P^2-M_\alpha^2)(Q^2-M_\beta^2)} \tag{3.5}$$

Here $\Psi_P([p_+^{(l)},\ \vec{p}_\perp^{(l)}])$ is the three-dimensional wave function of many-body system and is defined in the following way:

$$2(2\pi)^{-4}\delta(P_+-\sum_{l=1}^N p_+^{(l)})\delta^{(2)}(\vec{P}_\perp-\sum_{l=1}^N\vec{p}_\perp^{(l)})\Psi_P([p_+^{(l)},\vec{p}_\perp^{(l)}])=$$
$$=\int\prod_{l=1}^N dx_-^{(l)}d\vec{x}_\perp^{(l)}\exp\left[i\sum_{l=1}^N(\frac{1}{2}p_+^{(l)}x_-^{(l)}-\vec{p}_\perp^{(l)}\vec{x}_\perp^{(l)})\right]\times \tag{3.6}$$
$$\times\langle 0|\varphi_1(0,x_-^{(1)},\vec{x}_\perp^{(1)})\ldots\ \varphi_N(0,x_-^{(N)},\vec{x}_\perp^{(N)})|P,\alpha\rangle$$

Define the three-dimensional vertex function

$$\tilde{R}_{\mu\nu}([p_+^{(l)},\vec{p}_\perp^{(l)}];[q_+^{(l)},\vec{q}_\perp^{(l)}];z)=$$
$$=\int_0^{P_+}\prod_{l=1}^N dp_+^{(l)'}\delta(P_+-\sum^N p_+^{(l)'})\int\prod_{l=1}^N d\vec{p}_\perp^{(l)'}\delta^{(2)}(\vec{P}_\perp-\sum_{l=1}^N\vec{p}_\perp^{(l)'})\times$$
$$\times\int_0^{Q_+}\prod_{l=1}^N dq_+^{(l)'}\delta(Q_+-\sum_{l=1}^N q_+^{(l)'})\int\prod_{l=1}^N d\vec{q}_\perp^{(l)'}\delta^{(2)}(\vec{Q}_\perp-\sum_{l=1}^N\vec{q}_\perp^{(l)'})\times \tag{3.6}$$
$$\times\tilde{G}(P;[p_+^{(l)},\vec{p}_\perp^{(l)}];[p_+^{(l)'},\vec{p}_\perp^{(l)'}])\tilde{\Gamma}_{\mu\nu}([p_+^{(l)'},\vec{p}_\perp^{(l)'}];[q_+^{(l)'},\vec{q}_\perp^{(l)'}];z)\times$$
$$\times\tilde{G}(Q;[q_+^{(l)'},\vec{q}_\perp^{(l)'}];[q_+^{(l)},\vec{q}_\perp^{(l)}])$$

where $\tilde{G}$ is the three-dimensional Green function of N-body system, which has the following pole singularity in the vicinity of bound state $|P,\alpha\rangle$ with mass M_α and set of other quantum numbers α:

$$\tilde{G}(P;[p_+^{(l)},\vec{p}_\perp^{(l)}];[q_+^{(l)},\vec{q}_\perp^{(l)}])\Big|_{P^2\to M_\alpha^2} \simeq \tag{3.7}$$
$$\simeq \frac{i}{(2\pi)^4}\,\frac{\Psi_P([p_+^{(l)},\vec{p}_\perp^{(l)}])\,\Psi_Q^+([q_+^{(l)},\vec{q}_\perp^{(l)}])}{P^2-M_\alpha^2}$$

Taking into account the pole singularities of three-dimensional Green functions at $P^2\to M_\alpha^2$ and $Q^2\to M_\beta^2$ we can present the quantity $\tilde{R}_{\mu\nu}$ as follows:

$$\tilde{R}_{\mu\nu}([p_+^{(l)},\vec{p}_\perp^{(l)}];[q_+^{(l)},\vec{q}_\perp^{(l)}];z)\Big|_{\substack{P^2\to M_\alpha^2,\\ Q^2\to M_\beta^2}} \simeq$$
$$\simeq\left[\frac{i}{(2\pi)^4}\right]^2 \frac{\Psi_P([p_+^{(l)},\vec{p}_\perp^{(l)}])\,\Psi_Q^+([q_+^{(l)},\vec{q}_\perp^{(l)}])}{(P^2-M_\alpha^2)(Q^2-M_\beta^2)}\times$$
$$\times\int_0^{P_+}\prod_{l=1}^{N}dp_+^{(l)'}\,\delta(P_+-\sum_{l=1}^{N}p_+^{(l)'})\int\prod_{l=1}^{N}d\vec{p}_\perp^{(l)'}\,\delta^{(2)}(\vec{P}_\perp-\sum_{l=1}^{N}\vec{p}_\perp^{(l)'})\times \tag{3.8}$$
$$\times\int_0^{Q_+}\prod_{l=1}^{N}dq_+^{(l)'}\,\delta(Q_+-\sum_{l=1}^{N}q_+^{(l)'})\int\prod_{l=1}^{N}d\vec{q}_\perp^{(l)'}\,\delta^{(2)}(\vec{Q}_\perp-\sum_{l=1}^{N}\vec{q}_\perp^{(l)'})\times$$
$$\times\Psi_P^+([p_+^{(l)'},\vec{p}_\perp^{(l)'}])\,\tilde{\Gamma}_{\mu\nu}([p_+^{(l)'},\vec{p}_\perp^{(l)'}];[q_+^{(l)'},\vec{q}_\perp^{(l)'}];z)\,\Psi_Q([q_+^{(l)'},\vec{q}_\perp^{(l)'}])$$

Comparing this expression with (3.5), we get the following expression for the matrix element of T-product of currents:

$$\langle P,\alpha|T(J_\mu(z)J_\nu(0))|Q,\beta\rangle =$$

$$= \int_0^{P_+}\prod_{i=1}^{N} dp_+^{(i)}\,\delta(P_+-\sum_{i=1}^{N} p_+^{(i)})\int\prod_{i=1}^{N} d\vec{p}_\perp^{(i)}\,\delta^{(2)}(\vec{P}_\perp-\sum_{i=1}^{N}\vec{p}_\perp^{(i)})\times$$
$$\times\int_0^{Q_+}\prod_{i=1}^{N} dq_+^{(i)}\,\delta(Q_+-\sum_{i=1}^{N} q_+^{(i)})\int\prod_{i=1}^{N} d\vec{q}_\perp^{(i)}\,\delta^{(2)}(\vec{Q}_\perp-\sum_{i=1}^{N}\vec{q}_\perp^{(i)})\times \tag{3.9}$$
$$\times\Psi_P^+([p_+^{(i)},\vec{p}_\perp^{(i)}])\,\tilde{\Gamma}_{\mu\nu}([p_+^{(i)},\vec{p}_\perp^{(i)}];[q_+^{(i)},\vec{q}_\perp^{(i)}];z)\,\Psi_Q([q_+^{(i)},\vec{q}_\perp^{(i)}])$$

Fourier transform of this matrix element defines the amplitude of virtual Compton scattering of photon with space-like momentum q_μ on the hadron with momentum P_μ:

$$T_{\mu\nu}(P,q)=i\int d^4z\, e^{iqz}\,\langle P,\alpha|T(J_\mu(z)J_\nu(0))|P,\alpha\rangle =$$

$$= i\int_0^{P_+}\prod_{i=1}^{N} dp_+^{(i)}\,\delta(P_+-\sum_{i=1}^{N} p_+^{(i)})\int\prod_{i=1}^{N} d\vec{p}_\perp^{(i)}\,\delta^{(2)}(\vec{P}_\perp-\sum_{i=1}^{N}\vec{p}_\perp^{(i)})\times$$
$$\times\int_0^{Q_+}\prod_{i=1}^{N} dq_+^{(i)}\,\delta(Q_+-\sum_{i=1}^{N} q_+^{(i)})\int\prod_{i=1}^{N} d\vec{q}_\perp^{(i)}\,\delta^{(2)}(\vec{Q}_\perp-\sum_{i=1}^{N}\vec{q}_\perp^{(i)})\times \tag{3.10}$$
$$\times\Psi_P^+([p_+^{(i)},\vec{p}_\perp^{(i)}])\int d^4z\, e^{iqz}\,\tilde{\Gamma}_{\mu\nu}([p_+^{(i)},\vec{p}_\perp^{(i)}];[q_+^{(i)},\vec{q}_\perp^{(i)}];z)\,\Psi_Q([q_+^{(i)},\vec{q}_\perp^{(i)}])$$

According to optical theorem the tensor $W_{\mu\nu}$, which defines the hadronic part of deep inelastic lepton-hadron scattering cross section is related to the imaginary part of the amplitude of the zero angle virtual Compton scattering in the following way:

$$W_{\mu\nu}(P,q)=\sum_\alpha\int d^4z\, e^{iqz}\langle P,\alpha|J_\mu(z)J_\nu(0)|P,\alpha\rangle=\frac{1}{2\pi}\mathrm{Im}\,T_{\mu\nu}(P,q) \tag{3.11}$$

Taking into account the current conservation, the tensor $W_{\mu\nu}$ can be expressed via two invariant structure functions W_1 and W_2:

$$W_{\mu\nu}(P,q)=\left(-g_{\mu\nu}+\frac{q_\mu q_\nu}{q^2}\right)W_1(q^2,\nu)+$$
$$+\frac{1}{M^2}\left(P_\mu-\frac{Pq}{q^2}q_\mu\right)\left(P_\nu-\frac{Pq}{q^2}q_\nu\right)W_2(q^2,\nu) \tag{3.12}$$

where $M\nu=Pq$, M **is the hadron mass.**

Thus, using Equations (3.10)-(3.12) **one can express the structure functions of deep inelastic lepton-hadron scattering in terms of relativistic many-body wave functions, describing the internal motion of partons inside hadron, and the "two-photon" vertex function:**

$$\tilde{\Gamma}_{\mu\nu}([p_+^{(i)},\vec{p}_\perp^{(i)}];[q_+^{(i)},\vec{q}_\perp^{(i)}];q)=$$
$$=\int d^4z\, e^{iqz}\,\tilde{\Gamma}_{\mu\nu}([p_+^{(i)},\vec{p}_\perp^{(i)}];[q_+^{(i)},\vec{q}_\perp^{(i)}];z)=$$
$$=\int d^4z\, e^{iqz}\int_0^{P_+}\prod_{i=1}^{N}dp_+^{(i)'}\,\delta\Big(P_+-\sum_{i=1}^{N}p_+^{(i)'}\Big)\int\prod_{i=1}^{N}d\vec{p}_\perp^{(i)'}\,\delta^{(2)}\Big(\vec{P}_\perp-\sum_{i=1}^{N}\vec{p}_\perp^{(i)'}\Big)\times$$
$$\times\int_0^{Q_+}\prod_{i=1}^{N}dq_+^{(i)'}\,\delta\Big(Q_+-\sum_{i=1}^{N}q_+^{(i)'}\Big)\int\prod_{i=1}^{N}d\vec{q}_\perp^{(i)'}\,\delta^{(2)}\Big(\vec{Q}_\perp-\sum_{i=1}^{N}\vec{q}_\perp^{(i)'}\Big)\times \tag{3.13}$$
$$\times\tilde{G}^{-1}(P;[p_+^{(i)},\vec{p}_\perp^{(i)}];[p_+^{(i)'},\vec{p}_\perp^{(i)'}])\times$$
$$\times[\widetilde{G\Gamma_{\mu\nu}G}]([p_+^{(i)'},\vec{p}_\perp^{(i)'}];[q_+^{(i)'},\vec{q}_\perp^{(i)'}])\tilde{G}^{-1}(Q;[q_+^{(i)'},\vec{q}_\perp^{(i)'}];[q_+^{(i)},\vec{q}_\perp^{(i)}])$$

Here wave line over the $G\Gamma_{\mu\nu}G$ **denotes the integration over the "-" components of 4-momenta** $p^{(i)'}$ **and** $q^{(i)'}$.

2. Lowest Order in the Electromagnetic Interaction

The "two-photon" vertex operator $\tilde{\Gamma}_{\mu\nu}$ can be constructed using methods of perturbation theory and expanding the functions $\tilde{G}^{-1}$ and $\tilde{R}_{\mu\nu}=\widetilde{G\Gamma_{\mu\nu}G}$ in the series in coupling constant. In lowest order we obtain:

$$\tilde{\Gamma}^{(0)}_{\mu\nu}=[\tilde{G}^{(0)}]^{-1}[\widetilde{G^{(0)}\Gamma^{(0)}_{\mu\nu}G^{(0)}}][\tilde{G}^{(0)}]^{-1} \tag{3.15}$$

where multiplication is understood as integration over all "+" and "⊥" components of 4-momenta $p^{(i)'}$ and $q^{(i)'}$. $\tilde{G}^{(0)}$ is the three-dimensional Green function of N free particles.

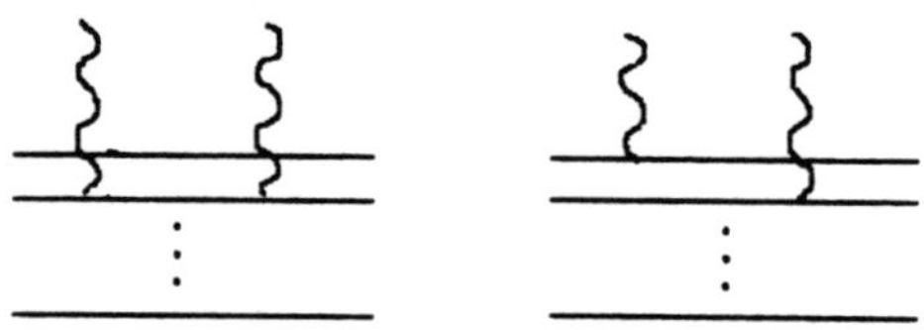

a b

Fig. 6.

In the lowest order two types of diagrams, shown in Fig. 6 will contribute in $\Gamma^{(0)}_{\mu\nu}$. The contribution of i-th diagram of type a) is equal to

$$\tilde{\Gamma}^{(0)}_{\mu\nu,i}([p^{(i)}_{+},\vec{p}^{\,(i)}_{\perp}];[q^{(i)}_{+},\vec{q}^{\,(i)}_{\perp}];q)=$$

$$=\frac{z_i^2}{2i(4\pi)^{N-1}}\Big(\prod_{i=1}^{N}\delta(p^{(i)}_{+}-q^{(i)}_{+})\,\delta^{(2)}(\vec{p}^{\,(i)}_{\perp}-\vec{q}^{\,(i)}_{\perp})\Big)\Big(\prod_{i=1}^{N}q^{(i)}_{+}\Big)\frac{1}{p^{(i)}_{+}(p^{(i)}_{+}+q^{(i)}_{+})}\times$$

$$\times\frac{(2\bar{p}^{(i)}+q)_{\mu}(2\bar{p}^{(i)}+q)_{\nu}}{P_{-}-\sum_{j=1}^{N}\frac{\vec{p}^{\,(j)2}_{\perp}+m_j^2}{p^{(j)}_{+}}+q_{-}+\frac{\vec{p}^{\,(i)2}_{\perp}+m_i^2}{p^{(i)}_{+}}+\frac{(\vec{p}^{\,(i)}_{\perp}+\vec{q}^{\,(i)}_{\perp})^2+m_i^2}{p^{(i)}_{+}+q^{(i)}_{+}}+i\varepsilon} \tag{3.16}$$

The contribution of diagrams of type b) is equal to:

$$\tilde{\Gamma}^{(o)}_{\mu\nu,b}([p_+^{(l)},\vec{p}_\perp^{(l)}];[q_+^{(l)},\vec{q}_\perp^{(l)}];q)=\frac{z_i z_j}{2i(4\pi)^{N-1}}\left(\prod_{l=1}^{N}q_+^{(l)}\right)\frac{1}{q_+^{(l)}q_+^{(j)}}\times$$

$$\times\,\delta(p_+^{(l)}-q_+-q_+^{(l)})\,\delta^{(2)}(\vec{p}_\perp^{(l)}-\vec{q}_\perp-\vec{q}_\perp^{(l)})\,\delta(p_+^{(j)}-q_+-q_+^{(j)})\,\delta^{(2)}(\vec{p}_\perp^{(j)}-\vec{q}_\perp-\vec{q}_\perp^{(j)})\times$$

$$\times\prod_{k=1}^{N}{}^{i,j}\,\delta(p_+^{(k)}-q_+^{(k)})\,\delta^{(2)}(\vec{p}_\perp^{(k)}-\vec{q}_\perp^{(k)})\frac{(2\overline{p}^{(l)}+q)_\mu(2\overline{p}^{(j)}+q)_\nu}{q_--\dfrac{\vec{p}_\perp^{(l)2}+m_l^2}{p_+^{(l)}}-\dfrac{\vec{q}_\perp^{(l)2}+m_l^2}{q_+^{(l)}}+i\varepsilon}\times \tag{3.17}$$

$$\times\left[\frac{Q_--\sum\limits_{j=1}^{N}\dfrac{\vec{q}_\perp^{(j)2}+m_j^2}{q_+^{(j)}}+i\varepsilon}{Q_--\sum\limits_{j=1}^{N}\dfrac{\vec{q}_\perp^{(j)2}+m_j^2}{q_+^{(j)}}-q_--\dfrac{\vec{p}_\perp^{(l)2}+m_l^2}{p_+^{(l)}}+\dfrac{\vec{q}_\perp^{(l)2}+m_l^2}{q_+^{(l)}}+i\varepsilon}-\right.$$

$$\left.-\frac{P_--\sum\limits_{j=1}^{N}\dfrac{\vec{p}_\perp^{(j)2}+m_j^2}{p_+^{(j)}}+i\varepsilon}{P_--\sum\limits_{j=1}^{N}\dfrac{\vec{p}_\perp^{(j)2}+m_j^2}{p_+^{(j)}}+q_--\dfrac{\vec{p}_\perp^{(l)2}+m_l^2}{p_+^{(l)}}-\dfrac{\vec{q}_\perp^{(l)2}+m_l^2}{q_+^{(l)}}+i\varepsilon}\right]$$

Assuming that the partons constituting hadron are on mass shell and neglecting small terms of order $P_--\sum_{l=1}^{N}(\vec{p}_\perp^{(l)2}+m_l^2)/p_+^{(l)}$, we obtain that diagrams of type b) do not give contribution to $\tilde{\Gamma}^{(o)}_{\mu\nu}$ and contribution of i-th diagram of type a) is equal to

$$\tilde{\Gamma}^{(o)}_{\mu\nu,i}([p_+^{(l)},\vec{p}_\perp^{(l)}];[q_+^{(l)},\vec{q}_\perp^{(l)}];q)=$$

$$=\frac{z_i^2}{2i(4\pi)^{N-1}}\,\frac{(2\overline{p}^{(i)}+q)_\mu(2\overline{p}^{(i)}+q)_\nu}{p_+^{(i)}[(\overline{p}^{(i)}+q)^2-m_i^2+i\varepsilon]}\prod_{l=1}^{N}p_+^{(l)}\delta(p_+^{(l)}-q_+^{(l)})\,\delta^{(2)}(\vec{p}_\perp^{(l)}-\vec{q}_\perp^{(l)}) \tag{3.18}$$

where z_i is the electric charge of i-th parton, and $\overline{p}^{(i)}$ is the momentum of parton on the mass shell:

$$\bar{p}^{(i)}=\left(\frac{\vec{p}_{\perp}^{(i)2}+m_i^2}{p_+^{(i)}},\ p_+^{(i)},\ \vec{p}_{\perp}^{(i)}\right).$$

Summing the contributions of all diagrams of type a), inserting into the expression (3.10) for virtual Compton scattering amplitude and extracting the imaginary part of the expression obtained, we get for tensor $W_{\mu\nu}$:

$$W_{\mu\nu}(P,q)=\frac{1}{4(4\pi)^{N-1}}\int_0^{P_+}\prod_{i=1}^{N}dp_+^{(i)}\ \delta(P_+-\sum_{i=1}^{N}p_+^{(i)})\times$$

$$\times\int\prod_{i=1}^{N}d\vec{p}_{\perp}^{(i)}\ \delta^{(2)}(\vec{P}_{\perp}-\sum_{i=1}^{N}\vec{p}_{\perp}^{(i)})\left(\prod_{i=1}^{N}p_+^{(i)}\right)\left|\Psi_P([p_+^{(i)},\vec{p}_{\perp}^{(i)}])\right|^2\times \tag{3.19}$$

$$\times\sum_{i=1}^{N}\left\{\frac{z_i^2}{p_+^{(i)}}(2\bar{p}^{(i)}+q)_{\mu}(2\bar{p}^{(i)}+q)_{\nu}\,\delta[(\bar{p}^{(i)}+q)^2-m_i^2]\right\}$$

Introducing the wave functions $\Phi_P([x^{(i)},\vec{p}_{\perp}^{(i)}])$ (see the Eq. (1.41)), we can rewrite the tensor $W_{\mu\nu}$ in the form:

$$W_{\mu\nu}(P,q)=\frac{1}{4(4\pi i)^{N-1}}\int_0^1\prod_{i=1}^{N}\frac{dx^{(i)}}{x^{(i)}}\delta(1-\sum_{i=1}^{N}x^{(i)})\times$$

$$\times\int\prod_{i=1}^{N}d\vec{p}_{\perp}^{(i)}\delta^{(2)}(\vec{P}_{\perp}-\sum_{i=1}^{N}\vec{p}_{\perp}^{(i)})\left|\Phi_P([x^{(i)},\vec{p}_{\perp}^{(i)}])\right|^2\times \tag{3.20}$$

$$\times\sum_{i=1}^{N}\left\{\frac{z_i^2}{x^{(i)}}(2\bar{p}^{(i)}+q)_{\mu}(2\bar{p}^{(i)}+q)_{\nu}\,\delta[(\bar{p}^{(i)}+q)^2-m_i^2]\right\}$$

Expressions for the structure functions W_1 and W_2 can be obtained multiplicating the tensor $W_{\mu\nu}$ by the "projection" operators $L_{\mu\nu}^{(i)}$:

$$L_{\mu\nu}^{(i)}W_{\mu\nu}=W_i\ ;\quad i=1,2 \tag{3.21}$$

Using the following expressions for these "projection" operators

$$L^{(1)}_{\mu\nu}=\frac{1}{2}\Big[-g_{\mu\nu}+\frac{P_\mu P_\nu}{M^2(1-\nu^2/q^2)}\Big] \tag{3.22a}$$

$$L^{(2)}_{\mu\nu}=\frac{1}{2(1-\nu^2/q^2)}\Big[-g_{\mu\nu}+\frac{3P_\mu P_\nu}{M^2(1-\nu^2/q^2)}\Big] \tag{3.22b}$$

for the structure functions W_1 and νW_2 we obtain:

$$W_1(q^2,\nu)=\frac{1}{8(4\pi)^{N-1}}\int_0^1\prod_{i=1}^{N}\frac{dx^{(i)}}{x^{(i)}}\delta(1-\sum_{i=1}^{N}x^{(i)})\times$$
$$\times\int\prod_{i=1}^{N}d\vec{p}_\perp^{(i)}\delta^{(2)}(\vec{P}_\perp-\sum_{i=1}^{N}\vec{p}_\perp^{(i)})\left|\Phi_P([x^{(i)},\vec{p}_\perp^{(i)}])\right|^2\times \tag{3.23a}$$
$$\times\sum_{i=1}^{N}\Big\{\frac{z_i^2}{x^{(i)}}\Big[\frac{(M\nu+2\bar{p}^{(i)}P)^2}{M^2(1-\nu^2/q^2)}-(4m_i^2-q^2)\Big]\delta(q^2+2\bar{p}^{(i)}q)\Big\}$$

$$\nu W_2(q^2,\nu)=\frac{1}{8(4\pi)^{N-1}}\frac{\nu}{(1-\nu^2/q^2)}\int_0^1\prod_{i=1}^{N}\frac{dx^{(i)}}{x^{(i)}}\delta(1-\sum_{i=1}^{N}x^{(i)})\times$$
$$\times\int\prod_{i=1}^{N}d\vec{p}_\perp^{(i)}\delta^{(2)}(\vec{P}_\perp-\sum_{i=1}^{N}\vec{p}_\perp^{(i)})\left|\Phi_P([x^{(i)},\vec{p}_\perp^{(i)}])\right|^2\times \tag{3.23b}$$
$$\times\sum_{i=1}^{N}\Big\{\frac{z_i^2}{x^{(i)}}\Big[\frac{3(M\nu+2\bar{p}^{(i)}P)^2}{M^2(1-\nu^2/q^2)}-(4m_i^2-q^2)\Big]\delta(q^2+2\bar{p}^{(i)}q)\Big\}$$

For further consideration we proceed to the frame, where the virtual photon and hadron are moving along the z axis:

$$P=(P_-,P_+,\vec{0}_\perp)\ ,\qquad q=(q_-,q_+,\vec{0}_\perp)$$

In this frame

$$2\bar{p}^{(i)}P=x^{(i)}M^2+\frac{\vec{p}_\perp^{(i)2}+m_i^2}{x^{(i)}}$$

and δ-function in (3.23) can be rewritten in the form:

$$\delta(q^2+2\bar{p}^{(i)}P)=\frac{1}{\xi}\delta\Big[\frac{\vec{p}_{\perp}^{(i)2}+m_i^2}{x^{(i)}}+\frac{Q^2(\xi-x^{(i)})}{\xi^2}\Big] \tag{3.24}$$

Here we have introduced the variables $Q^2=-q^2$ and

$$\xi=-\frac{q_{+}}{P_{+}}=\frac{Q^2}{M(\nu+\sqrt{\nu^2+Q^2})} \tag{3.25}$$

Scale properties of structure functions with respect of variable ξ are discussed in several papers [82, 84-87]. The Nachtman variable ξ is the generalization of usual Bjorken variable x_B taking into account the hadron mass and is related to x_B by the following relation:

$$\xi=\frac{2x_B}{1+\sqrt{1+\frac{4Mx_B}{Q^2}}} \tag{3.26}$$

Expressing the kinematic variable ν via the variables Q^2 and ξ we obtain the following expressions for structure functions:

$$W_1(Q^2,\xi)=\frac{1}{2(4\pi)^{N-1}\xi}\int_0^1\prod_{i=1}^{N}\frac{dx^{(i)}}{x^{(i)}}\delta(1-\sum_{i=1}^{N}x^{(i)})\times$$

$$\times\int\prod_{i=1}^{N}d\vec{p}_{\perp}^{(i)}\delta^{(2)}(\vec{P}_{\perp}-\sum_{i=1}^{N}\vec{p}_{\perp}^{(i)})\,|\Phi_P([x^{(i)},\vec{p}_{\perp}^{(i)}])|^2\times \tag{3.27a}$$

$$\times\sum_{i=1}^{N}\Big\{z_i^2\Big[\frac{Q^2x^{(i)}(x^{(i)}-\xi)}{\xi^2}-m_i^2\Big]\delta\Big[\vec{p}_{\perp}^{(i)2}+m_i^2-\frac{Q^2x^{(i)}(x^{(i)}-\xi)}{\xi^2}\Big]\Big\}$$

$$\nu W_2(Q^2,\xi)=\frac{MQ^2}{2(4\pi)^{N-1}\xi^2}\frac{(Q^2/\xi^2-M^2)}{(Q^2/\xi^2+M^2)^2}\int_0^1\prod_{i=1}^N\frac{dx^{(i)}}{x^{(i)}}\delta(1-\sum_{i=1}^N x^{(i)})\times$$

$$\times\int\prod_{i=1}^N d\vec{p}_\perp^{\,(i)}\delta^{(2)}(\vec{P}_\perp-\sum_{i=1}^N\vec{p}_\perp^{\,(i)})\,|\Phi_P([x^{(i)},\vec{p}_\perp^{\,(i)}])|^2\times \tag{3.27b}$$

$$\times\sum_{i=1}^N\Big\{z_i^2\Big[\frac{6Q^2x^{(i)}(x^{(i)}-\xi)}{\xi^2}+Q^2-2m_i^2\Big]\delta\Big[\vec{p}_\perp^{\,(i)2}+m_i^2-\frac{Q^2x^{(i)}(x^{(i)}-\xi)}{\xi^2}\Big]\Big\}$$

If one neglects the masses and transverse momenta of partons $(m_i^2\ll Q^2,\ \vec{p}_\perp^{\,(i)2}\ll Q^2)$, the δ-function takes the form

$$\delta(2\bar{p}^{(i)}q-Q^2)=\frac{\xi}{Q^2}\delta(x^{(i)}-\xi)$$

Then the structure function W_1 vanishes and for the structure function W_2 we obtain

$$\nu W_2(Q^2,\xi)=\frac{MQ^2}{2(4\pi)^{N-1}\xi}\frac{(Q^2/\xi^2-M^2)}{(Q^2/\xi^2+M^2)^2}\int_0^1\prod_{i=1}^N\frac{dx^{(i)}}{x^{(i)}}\delta(1-\sum_{i=1}^N x^{(i)})\times$$

$$\times\int\prod_{i=1}^N d\vec{p}_\perp^{\,(i)}\delta^{(2)}(\sum_{i=1}^N\vec{p}_\perp^{\,(i)})\,|\Phi_P([x^{(i)},\vec{p}_\perp^{\,(i)}])|^2\sum_{i=1}^N z_i^2\delta(x^{(i)}-\xi) \tag{3.28}$$

In the asymptotic limit $(\nu, Q^2\gg M^2,\ x_B$ is fixed) the variable ξ coinsides with the Bjorken variable x_B and we obtain that in this limit the structure function νW_2 is scale-invariant with respect to the variable x_B:

$$\nu W_2(x_B)=\frac{Mx_B}{2(4\pi)^{N-1}}\int_0^1\prod_{i=1}^N\frac{dx^{(i)}}{x^{(i)}}\delta(1-\sum_{i=1}^N x^{(i)})\times$$

$$\times\int\prod_{i=1}^N d\vec{p}_\perp^{\,(i)}\delta^{(2)}(\sum_{i=1}^N\vec{p}_\perp^{\,(i)})\,|\Phi_P([x^{(i)},\vec{p}_\perp^{\,(i)}])|^2\sum_{i=1}^N z_i^2\delta(x^{(i)}-x_B) \tag{3.29}$$

Normalization condition for wave function $\Phi([x^{(i)},\vec{p}_\perp^{\,(i)}])$ is of the form

$$
iP_+\int_0^1 \prod_{l=1}^N \frac{dx^{(l)}}{x^{(l)}}\,\delta(1-\sum_{l=1}^N x^{(l)}) \int \prod_{l=1}^N d\vec{p}_\perp^{\,(l)}\,\delta^{(2)}(\vec{P}_\perp - \sum_{l=1}^N \vec{p}_\perp^{\,(l)}) \times
$$

$$
\times \int_0^1 \prod_{l=1}^N \frac{dx^{(l)'}}{x^{(l)'}}\,\delta(1-\sum_{l=1}^N x^{(l)'}) \int \prod_{l=1}^N d\vec{p}_\perp^{\,(l)'}\,\delta^{(2)}(\vec{P}_\perp - \sum_{l=1}^N \vec{p}_\perp^{\,(l)'}) \times \tag{3.30}
$$

$$
\times\, \Phi_P([x^{(l)},\vec{p}_\perp^{\,(l)}])\, \frac{\partial \tilde{G}^{-1}(P;[p_+^{(l)},\vec{p}_\perp^{\,(l)}];[p_+^{(l)'},\vec{p}_\perp^{\,(l)'}])}{\partial P^2}\, \Phi_P([x^{(l)'},\vec{p}_\perp^{\,(l)'}]) = 1
$$

Assuming that the interaction kernel does not depend on the total energy and using the explicit expression for the Green function of N free particles we obtain the following sum rule:

$$
\int_0^1 \frac{\nu W_2(x_B)}{M x_B}\, dx_B = \sum_{l=1}^N z_l^2 . \tag{3.31}
$$

3. Model Parametrisations of Wave Functions

Consider now the case, when the hadron consists of two constituents. This case corresponds to meson, which consists of quark and antiquark. We will neglect contributions of gluons and quark-antiquark sea.

Expressions for structure functions in the case $N=2$ have the form:

$$
W_1(Q^2,\xi) = \frac{z_1^2 + z_2^2}{8\pi\xi} \int_a^1 \frac{dx}{x(1-x)} \int d\vec{p}_\perp\, |\Phi_P(x,\vec{p}_\perp)|^2 \times
$$

$$
\times \left[\frac{Q^2 x(x-\xi)}{\xi^2} - m^2 \right] \delta\left[\vec{p}_\perp^{\,2} + m^2 - \frac{Q^2 x(x-\xi)}{\xi^2} \right] \tag{3.32a}
$$

$$\nu W_2(Q^2,\xi)=\frac{z_1^2+z_2^2}{8\pi\xi^2}MQ^2\frac{Q^2/\xi^2-M^2}{(Q^2/\xi^2+M^2)^2}\times$$

$$\times\int_a^1\frac{dx}{x(1-x)}\int d\vec{p}_\perp\,|\Phi_P(x,\vec{p}_\perp)|^2\times \tag{3.32b}$$

$$\times\left[\frac{6Q^2x(x-\xi)}{\xi^2}+Q^2-2m^2\right]\delta\left[\vec{p}_\perp^{\,2}+m^2-\frac{Q^2x(x-\xi)}{\xi^2}\right]$$

Here we assume that masses of constituents are equal to each other: $m_1=m_2=m$.

In (3.32a,b) the limit of integration over x is defined from the δ-function and equals to:

$$a=\frac{\xi}{2}\left(1+\sqrt{1+\frac{4m^2}{Q^2}}\right) \tag{3.33}$$

Neglecting masses and transverse momenta of quarks we obtain that the structure function W_1 vanishes and the structure function νW_2 takes the following form:

$$\nu W_2(Q^2,\xi)=\frac{(z_1^2+z_2^2)MQ^2}{8\pi\xi^2(1-\xi)}\frac{Q^2/\xi^2-M^2}{(Q^2/\xi^2+M^2)^2}\int d\vec{p}_\perp|\Phi_P(\xi,\vec{p}_\perp)|^2 \tag{3.34}$$

If we choose the following parametrization for wave function Φ_P

$$\Phi_P(x,\vec{p}_\perp)=C\left[\frac{\vec{p}_\perp^{\,2}+m^2}{x(1-x)}-\alpha\right]^{-n} \tag{3.35}$$

then for the structure function νW_2 we get

$$\nu W_2(Q^2,\xi)=\frac{(z_1^2+z_2^2)MQ^2}{8}\frac{Q^2/\xi^2-M^2}{(Q^2/\xi^2+M^2)^2}\frac{|C|^2\xi^{2n-1}(1-\xi)^{2n-1}}{(2n-1)[m^2-\alpha\xi(1-\xi)]^{2n-1}} \tag{3.36}$$

In the Bjorken limit $(Q^2 \gg m^2,\ \xi \to x_B)$ for $n=1$ we obtain [88]:

$$\nu W_2(x_B) = \frac{z_1^2 + z_2^2}{8} M\,|C|^2 \frac{x_B(1-x_B)}{m^2 - \alpha x_B(1-x_B)} \tag{3.37}$$

Consider now another parametrization of the wave function

$$\Phi_P(x, \vec{p}_\perp) = C \exp\left[-\beta \frac{\vec{p}_\perp^{\,2} + m^2}{x(1-x)}\right] \tag{3.38}$$

Inserting this wave function into (3.32a) and (3.32b) and neglecting quark masses $(m^2 \ll Q^2)$, for the structure functions we get:

$$\begin{aligned} W_1(Q^2,\xi) &= \frac{(z_1^2+z_2^2)Q^2(1-\xi)}{8\xi^3}|C|^2 \times \\ &\times \left\{-1-\left(1+2\beta\frac{Q^2}{\xi^2}\right)\exp\left(2\beta\frac{Q^2}{\xi^2}\right)\mathrm{Ei}\left(-2\beta\frac{Q^2}{\xi^2}\right)\right\} \end{aligned} \tag{3.39a}$$

$$\begin{aligned} \nu W_2(Q^2,\xi) &= \frac{6MQ^2}{\xi}\,\frac{Q^2/\xi^2 - M^2}{(Q^2/\xi^2+M^2)^2}\,W_1(Q^2,\xi) + \\ &+ \frac{(z_1^2+z_2^2)MQ^4}{4\xi^2}\,\frac{Q^2/\xi^2-M^2}{(Q^2/\xi^2+M^2)^2}|C|^2\left\{-\exp\left(2\beta\frac{Q^2}{\xi^2}\right)\mathrm{Ei}\left(-2\beta\frac{Q^2}{\xi^2}\right)\right\} \end{aligned} \tag{3.39b}$$

Here Ei is the integral exponent function.

Note that more realistic consideration needs inclusion of spin and color degrees of freedom of constituents.

4. Quark Degrees of Freedom in Nuclei and EMC-Effect

As has been mentioned in the Introduction recent investigations have led to the conclusion that the consideration of nuclei as the systems of quasi-independent nonrelativistic nucleons is incomplete and they require a relativistic description of the internal motion of nucleons in nuclei and taking into account quark degrees of freedom. These are first of all the prediction [35] and observation [36] of the cumulative production of particles in hadron-nucleus and nucleus-nucleus collisions and change of an exponential fall-off at small momentum transfers of form-factors of light nuclei to a power-law fall-off at large momentum transfers [89] according to the quark counting rules [67,68].

In this sense most impressive are the results of recent deep inelastic lepton-nucleus experiments [90-93] (the so-called EMC-effect). Different models are suggested for the explanation of this phenomenon (see, e.g., Refs. [94-107]). It seems that the above mentioned regularities are of the same nature and are determined by the possibility of formation of multiquark configurations (multiquark bags) in nuclei. Here we analyse the EMC-effect and show that the effect can be explained by taking into account the scattering on colourless multiquark configurations which are contained in the medium of spectator nucleons.

Consider deep inelastic scattering of charged leptons on a nucleus A. We shall assume that in the nucleus, together with nucleons (three-quark bags), with definite probabilities the configurations with six, nine, etc. quarks (see in this connection [8,20,21]) are formed and leptons interact with the nucleus by the exchange of virtual photons with quarks from these bags. Then the nucleus structure function can be represented by the sum:

$$F_2(x,Q^2) = \sum_{K=1}^{A} N(A,K) F_2^K(x,Q^2) \tag{3.40}$$

where F_2^K is the structure function of the nucleus A which contains a 3K-quark bag and (A-K) nucleons. The coefficients N(A,K) of these structure function have the meaning of the effective number of 3K-quark bags in the nucleus A and obey the normalization condition:

$$\sum_{K=1}^{A} K\,N(A,K) = A \tag{3.41}$$

We use the parametrization of N(A,K) in the form of the Bernoulli distribution:

$$N(A,K) = \frac{A!}{K!(A-K)!} p(A)^{K-1} [1-p(A)]^{A-K} \tag{3.42}$$

For the parameter p(A) determining the probability of a three-quark nucleon to get into a 3K-quark bag we consider two possibilities [8]:

1). p(A) is determined by the ratio of the bag and nucleus volumes:

$$p(A) = \frac{V_K}{V_A} = \frac{r_K^3}{R_A^3} = \frac{r_K^3}{(R_0 A^{1/3})^3} \sim A^{-1} \tag{3.43}$$

Taking the bag radius r_K close to the nucleon radius $r_K \approx 0{,}8$ fm and $R_0 = 1{,}4$ fm [108] we obtain $p(A) = 1{,}1867\,A^{-1}$.

2). p(A) is determined by the ratio of the bag and nucleus cross sections:

$$p(A) \sim \frac{r_K^2}{R_A^2} \sim A^{-2/3} \tag{3.44}$$

The coefficient of proportionality in (3.44) for this case has been obtained in Ref. [109] by fitting the data on production of π-mesons with large transverse momenta in proton-nucleus scattering, and we shall also use this parametrization $p(A)=0{,}085\,A^{-0{,}67}$.

In both cases $N(A,K)$ are rapidly decreasing functions of K and the main contribution to the structure function is given by first few terms of the sum (3.40).

We proceed now to the calculation of structure functions F_2^K. As is well-known, the structure function F_2 appears in the decomposition of the deep inelastic tensor $W_{\mu\nu}$ into the gauge-invariant structures. The tensor $W_{\mu\nu}$ itself is proportional to the imaginary part of the forward virtual Compton scattering amplitude:

$$T_{\mu\nu}(P_A,q)=i\int d^4z\,\exp(iqz)\langle P_A|T\{J_\mu(z)J_\nu(0)\}|P_A\rangle \qquad (3.45)$$

that can be expressed through the two-photon vertex function and relativistic wave functions of composite systems:

$$\begin{aligned}
T_{\mu\nu}(P_A,q)=i\int_0^{P_{A,+}}&\prod_{i=K+1}^{A} dp_+^{(i)}dP_{K,+}\int\prod_{i=K+1}^{A} d\vec{p}_\perp^{(i)}d\vec{P}_{K,\perp}\times\\
&\times\delta(P_{A,+}-P_{K,+}-\sum_{i=K+1}^{A}p_+^{(i)})\,\delta^{(2)}(\vec{P}_{A,\perp}-\vec{P}_{K,\perp}-\sum_{i=K+1}^{A}\vec{p}_\perp^{(i)})\times\\
&\times\int_0^{Q_{A,+}}\prod_{i=K+1}^{A} dq_+^{(i)}dQ_{K,+}\int\prod_{i=K+1}^{A} d\vec{q}_\perp^{(i)}d\vec{Q}_{K,\perp}\times \qquad (3.46)\\
&\times\delta(Q_{A,+}-Q_{K,+}-\sum_{i=K+1}^{A}q_+^{(i)})\,\delta^{(2)}(\vec{Q}_{A,\perp}-\vec{Q}_{K,\perp}-\sum_{i=K+1}^{A}\vec{q}_\perp^{(i)})\times\\
\times\Psi_{P_A}^{+}(P_{K,+},\vec{P}_{K,\perp},[p_+^{(i)},\vec{p}_\perp^{(i)}])\,&\Gamma_{\mu\nu}(P_{K,+},\vec{P}_{K,\perp},[p_+^{(i)},\vec{p}_\perp^{(i)}];q;[q_+^{(i)},\vec{q}_\perp^{(i)}],Q_{K,+},\vec{Q}_{K,\perp})\times\\
&\times\Psi_{Q_A}(Q_{K,+},\vec{Q}_{K,\perp},[q_+^{(i)},\vec{q}_\perp^{(i)}])
\end{aligned}$$

Here $\Gamma_{\mu\nu}$ is the two-photon vertex function, Ψ_{P_A} is the relativistic wave function of the composite system consisting of a 3K-quark bag and (A-K) nucleons. $P_{A,\mu}(\mu=0,1,2,3)$ is the 4-momentum of nucleus A, $P_{K,\mu}$ and $Q_{K,\mu}$ are the 4 momenta of the 3K-quark bag, $p_\mu^{(i)}$, $q_\mu^{(i)}$ are the 4-momenta of nucleons. The square brackets in the arguments of Ψ_{P_A} and $\Gamma_{\mu\nu}$ denote the sets of corresponding variables:

$$[p_+^{(i)}, \vec{p}_\perp^{(i)}] = p_+^{(K+1)}, \vec{p}_\perp^{(K+1)}; p_+^{(K+2)}, \vec{p}_\perp^{(K+2)}; \ldots ; p_+^{(A)}, \vec{p}_\perp^{(A)}. \qquad (3.47)$$

Represent now the relativistic wave function $\Psi_{P_A}(P_{K,+}, \vec{P}_{K,\perp}, [p_+^{(i)}, \vec{p}_\perp^{(i)}])$ in the form:

$$\Psi_{P_A}(P_{K,+}, \vec{P}_{K,\perp}, [p_+^{(i)}, \vec{p}_\perp^{(i)}]) = \Psi_{A-K}([p_{i,+}, \vec{p}_{i,\perp}]) \times$$
$$\times \int \prod_{j=1}^{3K} dk_{j,+}\, \delta(P_{K,+} - \sum_{j=1}^{3K} k_{j,+}) \int \prod_{j=1}^{3K} d\vec{k}_{j,\perp}\, \delta^{(2)}(\vec{P}_{K,\perp} - \sum_{j=1}^{3K} \vec{k}_{j,\perp})\, \Psi_{3K}([k_{j,+}, \vec{k}_{j,\perp}]) \qquad (3.48)$$

where $k_{j,+}$ and $\vec{k}_{j,\perp}$ are components of the quark momentum in the 3K-quark bag.

Calculating the function $\Gamma_{\mu\nu}$ in the lowest order in electromagnetic interaction, we get the following expression for the structure function F_2^K:

$$F_2^K(x, Q^2) = \frac{\pi}{(4\pi)^{A+2K}} \frac{M_A Q^2 (Q^2/\xi_A^2 - M_A^2)}{\xi_A^2 (Q^2/\xi_A^2 + M_A^2)^2} \times$$
$$\times \int_0^1 \prod_{i=K+1}^{A} \frac{dx_i}{x_i} Z_K\, dZ_K\, \delta(1 - Z_K - \sum_{i=K+1}^{A} x_i) \times$$
$$\times \int \prod_{i=K+1}^{A} d\vec{p}_{i,\perp}\, d\vec{P}_{K,\perp}\, \delta^{(2)}(\vec{P}_{K,\perp} + \sum_{i=K+1}^{A} \vec{p}_{i,\perp}) |\Phi_{A-K}([x_i, \vec{p}_{i,\perp}])|^2 \times$$
$$\times \sum_{j=1}^{3K} e_j^2 \int_0^1 \prod_{l=1}^{3K} \frac{dz_l}{z_l} \delta(1 - \sum_{l=1}^{3K} z_l) \times \qquad (3.49)$$

$$\times\int\prod_{l=1}^{3K}d\vec{k}_{l,\perp}\,\delta^{(2)}(\vec{P}_{K,\perp}-\sum_{l=1}^{3K}\vec{k}_{l,\perp})\,|\Phi_{3K}([z_l,\vec{k}_{l,\perp}])|^2\times$$

$$\times\left[Q^2-2m_j^2+\frac{6Q^2}{(\xi_A/Z_K)^2}z_j(z_j-\xi_A/Z_K)\right]\delta^{(2)}\left[\vec{k}_{j,\perp}^{\,2}+m_j^2-\frac{Q^2z_j(z_j-\xi_A/Z_K)}{(\xi_A/Z_K)^2}\right]$$

Here $q^2=-Q^2$ **is the** 4-**momentum transfer squared**. **The variable** ξ_A **is defined as**

$$\xi_A=\frac{2x_A}{1+(1+4M_A^2x_A^2/Q^2)^{1/2}}=z_jZ_K=\frac{k_{j,+}}{P_{A,+}}. \tag{3.50}$$

where $x_A=Q^2/2M_A\nu$, $\nu=(P_Aq)/M_A$. M_A **is the nucleus mass. The variable** x_A **varies in the interval** $0<x_A<1$ **and is related to the Bjorken variable** $x=Q^2/2M\nu$ **by** $x_A=(M/M_A)x$ (M **is the nucleon mass). It is evident that** $0<x<M_A/M\approx A$ **and**

$$\xi_A=\frac{M}{M_A}\xi\approx\frac{\xi}{A},\qquad \xi=\frac{2x}{1+(1+4M^2x^2/Q^2)^{1/2}}. \tag{3.51}$$

In (3.49) e_j **and** m_j **are the electric charge and the mass of an** j-**th quark, respectively.**

The wave functions $\Phi_{A-K}([x_l,\vec{p}_{l,\perp}])$ **and** $\Phi_{3K}([z_j,\vec{k}_{j,\perp}])$ **are related to the functions** $\Psi_{A-K}([\underline{p}_l])$ **and** $\Psi_{3K}([\underline{k}_j])$ **by the following formulas:**

$$\Phi_{A-K}([x_l,\vec{p}_{l,\perp}])=(P_{A,+})^{A-K}\Big(\prod_{l=K+1}^{A}x_l\Big)\Psi_{A-K}([\underline{p}_l]) \tag{3.52a}$$

$$\Phi_{3K}([z_j,\vec{k}_{j,\perp}])=(P_{K,+})^{3K-1}\Big(\prod_{j=1}^{3K}z_j\Big)\Psi_{3K}([\underline{k}_j]) \tag{3.52b}$$

The variables x_l **and** z_j **are defined as:**

$$x_l=\frac{p_{l,+}}{P_{A,+}},\quad 0<x_l<1,\quad \sum_{l=K+1}^{A}x_l=1-Z_K,$$

$$z_j=\frac{k_{j,+}}{P_{K,+}},\quad 0<z_j<1,\quad \sum_{j=1}^{3K}z_j=1.$$

The variable Z_K is the ratio of "+"-components of the 4-momenta of the 3K-quark bag and nucleus A, $\vec{p}_{l,\perp}$ and $\vec{k}_{j,\perp}$ are transverse momenta of the nucleons in the nuclei and of a quarks in the bag, respectively.

Let us choose now the wave functions $\Phi_{A-K}([x_l,\ \vec{p}_{l,\perp}])$ and $\Phi_{3K}([z_j,\vec{k}_{j,\perp}])$ in the following form:

$$\Phi_{A-K}([x_l,\vec{p}_{l,\perp}])\sim\exp\Big[-\alpha_A\sum_{l=K+1}^{A}\frac{\vec{p}_{l,\perp}^{\,2}+M_l^2}{x_l}\Big]\tag{3.53}$$

$$\Phi_{3K}([z_j,\vec{k}_{j,\perp}])\sim\exp\Big[-\beta_K\sum_{j=1}^{3K}\frac{\vec{k}_{j,\perp}^{\,2}+m_j^2}{z_j}\Big]\tag{3.54}$$

M_l are the nucleon masses, m_j are the quark masses.

In the deep inelastic limit ($Q^2\gg M_l^2,\ m_j^2$; $\xi_A\to x_A$; $\xi\to x$) the Q^2-dependence in the structure functions F_2^K disappears, and after the corresponding calculations from (3.49) for the quark distribution function $f^K(x)$, which is defined by the following relation

$$F_2^K(x)=\langle e_q^2\rangle\, x f^K(x)\tag{3.55}$$

($\langle e_q^2\rangle$ is the average value of quark charges squared), we obtain:

$$f^K(x)=\frac{3K}{A}\,\frac{I_K(x_A)}{\int_0^1 dx_A\ I_K(x_A)}\ ;\qquad x_A=\frac{x}{A},\quad K=1,2,\dots,A-1\tag{3.56}$$

where

$$I_K(x_A) = \int_{x_A}^{1} dZ_K \frac{(1-Z_K)^{A-K-1}}{1+(\beta_K/\alpha_A)(1-Z_K)}\left(1-\frac{x_A}{Z_K}\right)^{3K-2} =$$

$$= \frac{x_A(1-x_A)^{(A-K)+(3K-2)}}{1+(\beta_K/\alpha_A)(1-x_A)} B(A-K, 3K-1) \times \tag{3.57}$$

$$\times\, {}_2F_1\left(3K-1, A-K, 1, A+2K-1; 1-x_A; \frac{(1+\beta_K/\alpha_A)(1-x_A)}{1+(\beta_K/\alpha_A)(1-x_A)}\right)$$

Here B is the Euler beta-function, ${}_2F_1$ is the hypergeometric function of two variables (the Appel function).

A characteristic feature of structure function F_2^K is a possible existence in a nucleus of a superfast quark which in expreme situation takes all the momentum of nucleus.

The structure functions (3.55) - (3.57) are normalized to the number of quarks in the corresponding bag

$$\int_0^A dx\, f^K(x) = 3K \tag{3.58}$$

One can relate the parameters α_A and β_K of the relativistic wave functions $\Phi_{A-K}([x_i, \vec{p}_{i,\perp}])$ and $\Phi_{3K}([z_j, \vec{k}_{j,\perp}])$ to the radii of the nucleus A and 3K-quark bag, respectively

$$R_A^2 \sim 8A\alpha_A \tag{3.59a}$$

$$r_K^2 \sim 24K\beta_K \tag{3.59b}$$

As can be seen from (3.56) and (3.57) the structure functions F_2^K depend on the ratio

$$\frac{\beta_K}{\alpha_A} = \frac{A}{3K}\frac{r_K^2}{R_A^2} = \frac{r_K^2 A^{1/3}}{3R_0^2 K} \tag{3.60}$$

Choosing $r_K = 0.8$ fm, $R_0 = 1.4$ fm we get $\beta_K/\alpha_A = 0.109\, A^{1/3}/K$.

In the experiments [90-93] the ratio of structure functions $F_2(Fe)/F_2(D)$ is measured in the region $x < 1$ that corresponds to the region of small $x_A < 1/A$. It is evident from (3.55)-(3.56) and it can be checked by straightforward numerical calculations that the structure functions weakly depend on the parameter β_K/α_A in this region. That is why in the numerical calculations we have supposed that the ratio β_K/α_A is tha same for all bags. For the iron nucleus we have taken $(\beta_K/\alpha_A)_{Fe} = 0.1$. But for the deuteron (considering that the average distance between nucleons in the deuteron is somewhat larger than in other nuclei) we have chosen the value $(\beta_K/\alpha_A)_D = 0.05$.

Experimental data on the ratio $F_2(Fe)/F_2(D)$ and the curves calculated by the formulas (3.40), (3.42), (3.55)-(3.57) with two parametrizations for $p(A)$ are shown in Fig. 7. The curves reproduce rather well the experimental data in the region $x > 0.2$ and somewhat differ from the data in the region $x < 0.2$. This difference is probably due to the fact that we restrict our consideration by the valence quarks and do not take into account contributions of the sea quarks and gluons.

The analysis performed shows that deviation of the ratio $F_2(Fe)/F_2(D)$ from unity is due to a larger contribution of the multiquark configurations in the iron nucleus as compared with the deuteron. Note that according to (3.42) the probabilities of formation of six-, nine-, etc., quark configurations in heavy nuclei are larger than in light nuclei. In the deuteron the contribution of a six-quark state is only a small admixture to the contribution of a two-

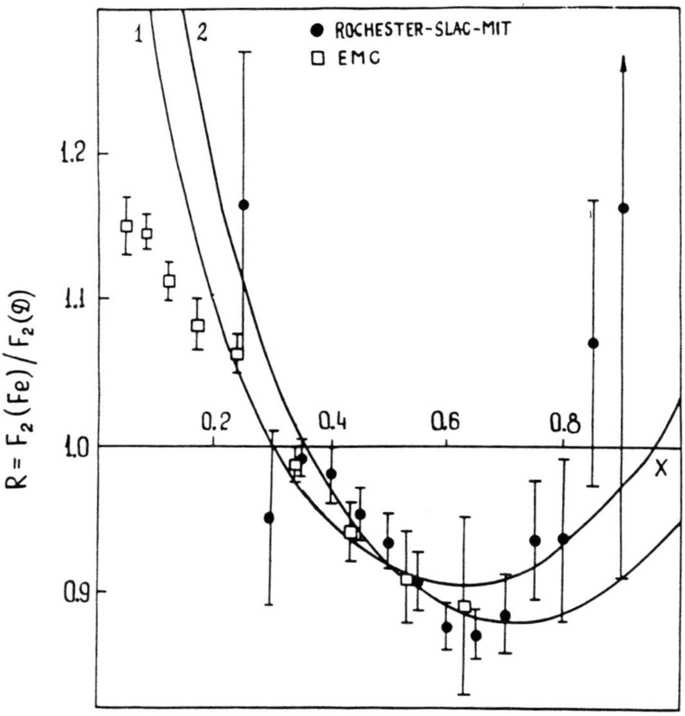

Fig. 7. The ratio of the structure functions of iron and deuteron. Curves 1 and 2 - calculations with parametrizations $P(A) \sim A^{-1}$ and $P(A) \sim A^{-2/3}$, respectively.

nucleon state. The deviation from unity of the ratio of the sum of two-nucleon and six-quark contributions to the contribution of a pure two-nucleon state does not exceed 5% for the deuteron in the whole region $0 < x < 2$ (in this connection see also [110, 111]).

In Refs. [16,18] the possibility was established for extracting information on the quark parton functions of nuclei from the data on cumulative pion production, and in Ref. [112] similarity in the x-behaviour of the ratio of pion production cross sections on different nuclei and the ratio of the deep inelastic structure functions of the same nuclei was pointed out. It should be noticed that the data on cumulative production allow us to investigate the region $x > 1$ (which is not yet reached in the experiments on deep inelastic lepton-nucleus scattering). We have calculated the ratios $F_2(Fe)/F_2(D)$, $F_2(Fe)/F_2(He)$ and $F_2(Fe)/F_2(Al)$ in the whole region of variable x $(0 < x < A)$. The curves of these calculations in the parametrization $p(A) \sim A^{-1}$ in the double logarithmic presentation are given in Fig. 8. A similarity in the behaviour of these curves and the ratios of cross sections of cumulative pion production is observed.

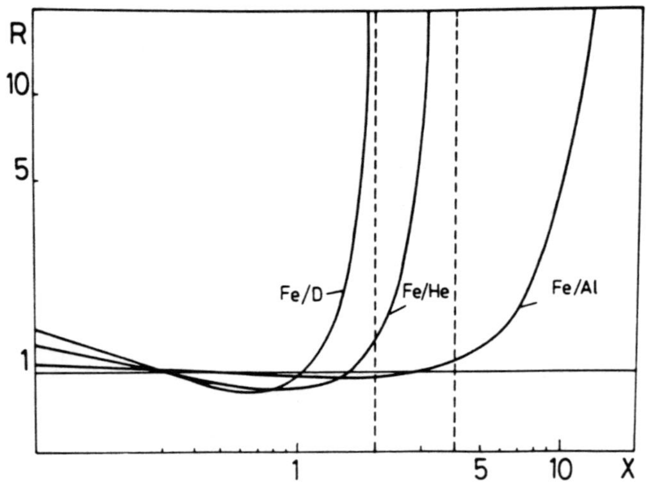

Fig. 8. The ratios of structure functions of different nuclei in the whole region of variable x $(0 < x < A)$

The analysis performed shows that the explanation of the observed deviation from unity of the ratio $F_2(Fe)/F_2(D)$ requires the consideration of multiquark configurations in nuclei. It seems very interesting to analyse the A-dependence of the ratio $F_2(A)/F_2(D)$ contained in (3.40), (3.42), (3.54)-(3.57) (there are the data on this dependence [92]) and experimental investigations of the deep inelastic lepton-nucleus scattering in the region, where the variable $x > 1$. (It seems that until now in the region $1 < x < 1.4$ only the structure function of ^{12}C nucleus has been measured [113]).

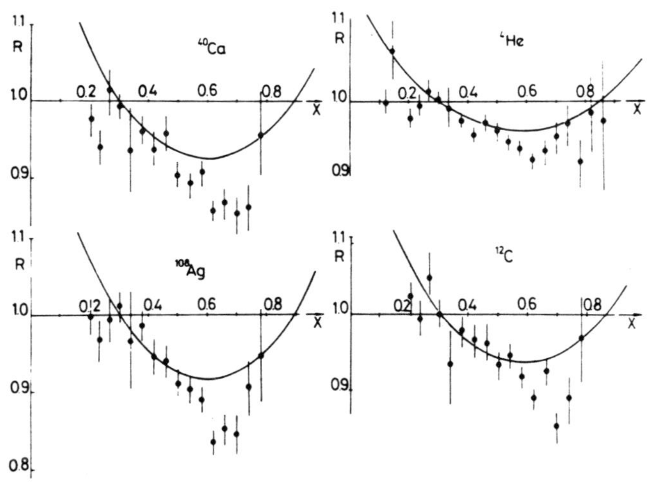

Fig. 9. The ratios of structure functions of ^4He, ^{12}C, ^{40}Ca, ^{108}Ag nuclei to structure function of deuteron versus variable x. Data from Ref. [92].

On Fig. 9 the theoretical curves on the ratio $F_2(A)/F_2(D)$ for several different nuclei and experimental data on deep inelastic scattering cross section ratio σ_A/σ_D are presented and good agreement is observed. More correct comparison of these quantities needs taking into account the contributions from structure function F_1 to the cross section. On Fig. 10 the A-dependence of these ratios is presented at various values of variable Q^2. The Fig. 11 shows the A-dependence of ratio $F_2(A)/F_2(Pb)$ at different values of x. Quantitevely these curves at $x > 1$ have the same behaviour as the data on cumulative pion production.

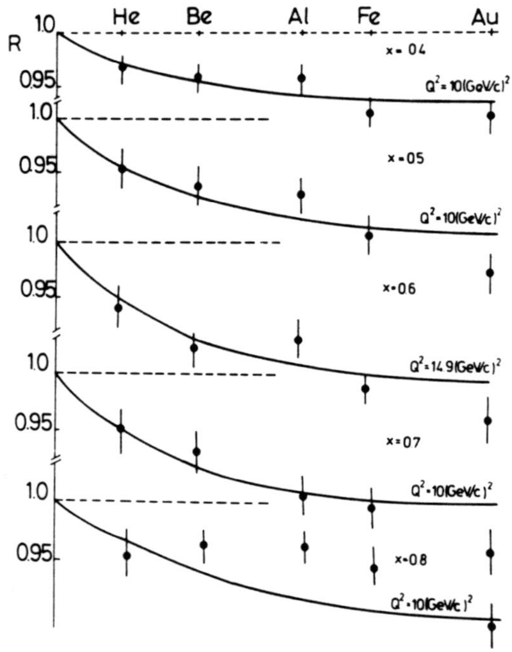

Fig. 10. The A-dependence of structure function ratios at various values of variable Q^2. Data from Ref. [92].

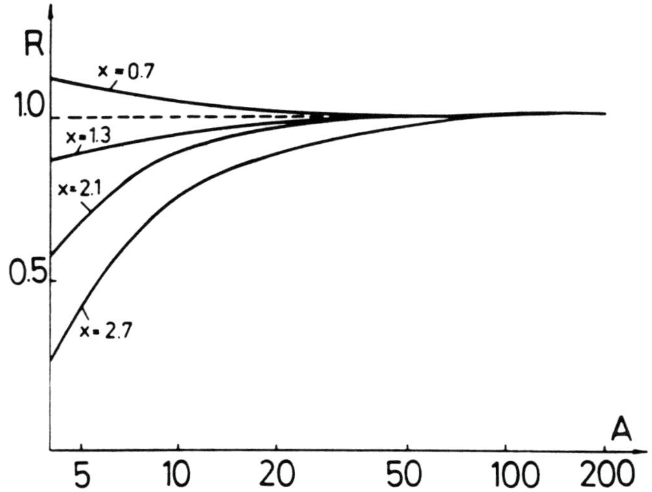

Fig. 11. The A-dependence of ratio $F_2(A)/F_2(Pb)$ at different values of variable x.

Note that in the region of very small x the experimental situation has been changed significantly [114, 115] and the problem of the theoretical explanation of this phenomenon is of special interest (see in this connection Ref. [116]).

Chapter IV

PROCESSES INVOLVING HIGH ENERGY NUCLEI AND THE PROBLEM OF RELATIVIZATION OF NUCLEAR WAVE FUNCTIONS

1. Scale-Invariant Parametrization of the Deuteron Relativistic Wave Function

In this Chapter we consider the relativization of nuclear wave functions. Let's start our consideration with the simplest case of a two-nucleon nucleus, i.e. deuteron. To construct a relativistic wave function of deuteron, we use the fact that in the framework of "light front" formalism the Lorenz-invariant combination $(\vec{p}_\perp^{\,2} + m^2)/x(1-x)$ (where $\vec{p}_\perp$ is the relative transverse momentum of the constituent , $x = 1/2 + (p_0 + p_z)/(P_0 + P_z)$, p_μ and P_μ are the relative 4-momentum of the internal motion of the constituents and the total 4-momentum of the composite system, respectively, m is the constituent mass) plays the role analogous to that of $\vec{p}^{\,2}$ - the square of three-dimensional relative momentum in nonrelativistic theory, which is invariant under three-dimensional rotations. The relativistic wave functions can be obtained from the corresponding nonrelativistic expressions by substitution [7,117,118]

$$\vec{p}^{\,2} \to \frac{\vec{p}_\perp^{\,2} + m^2}{x(1-x)}$$

and changing the nonrelativistic numerical parameters by the relativistic ones.

In this way one can obtain, for example, the relativistic analog of the

Hulten wave function:

$$\Phi_R(x,\vec{p}_\perp)=C_R\left[\frac{\vec{p}_\perp^{\,2}+m^2}{x(1-x)}-\alpha_R\right]^{-1}\left[\frac{\vec{p}_\perp^{\,2}+m^2}{x(1-x)}-\beta_R\right]^{-1} \qquad (4.1)$$

which is written in an arbitrary reference frame for the arbitrary momenta of the deuteron as a whole and arbitrary intrinsic momenta of its constituent nucleons. Note that relativization of other, more accurate, deuteron wave functions does not meet any principal difficulties. In particular, in what follows we shall use the relativistic analog of the Gartenhaus-Moravchik wave function. We shall not describe here other possible ways of relativization of the wave functions, which can be found in Refs. [119-125].

In the deuteron rest frame, when the momenta of internal motion of its constituents obey the condition $|\vec{p}|/m_N \ll 1$, the wave function (4.1) turns to the well-known nonrelativistic Hulten wave function

$$\Phi_{NR}(\vec{p})=C_{NR}(\vec{p}^{\,2}+\alpha_{NR}^2)^{-1}(\vec{p}^{\,2}+\beta_{NR}^2)^{-1} \qquad (4.2)$$

In (4.1) and (4.2) α_R, β_R and α_{NR} , β_{NR} are adjustable parameters of the relativistic and nonrelativistic wave functions, respectively, C_R and C_{NR} are normalization coefficients.

Neglecting the dependence of variable x on binding energy from the condition that the wave function Φ_R go over into the wave function Φ_{NR} in the nonrelativistic limit we obtain the following relation between the parameters

$$\alpha_R=4(m^2-\alpha_{NR}^2); \quad \beta_R=4(m^2-\beta_{NR}^2) \qquad (4.3)$$

From the normalization condition for wave function (4.2)

$$\int d\vec{p}\,|\Phi_{NR}(\vec{p})|^2=1 \qquad (4.4)$$

we get the following expression for normalization coefficient C_{NR}

$$C_{NR}=\frac{1}{\pi}(\alpha_{NR}+\beta_{NR})^{3/2}\alpha_{NR}^{1/2}\beta_{NR}^{1/2} \tag{4.5}$$

For the normalization of the wave function (4.1) it is necessary to know the form of all interactions within the two-particle bound state. Assuming, that the total interaction kernel does not depend on the total 4-momentum of the deuteron, we obtain the following normalization condition:

$$\int_0^1\frac{dx}{x(1-x)}\int d\vec{p}_\perp|\Phi_R(x,\vec{p}_\perp)|^2=8\pi \tag{4.6}$$

Inserting into this normalization condition the wave function (4.1) we get the following expression for normalization coefficient C_R

$$C_R=2^{3/2}(\alpha_R-\beta_R)\left[f(\alpha_R,\beta_R)+f(\beta_R,\alpha_R)\right]^{-1/2} \tag{4.7}$$

where

$$f(\alpha_R,\beta_R)=\frac{4\left[m^2(\alpha_R-\beta_R)+\alpha_R(4m^2-\alpha_R)\right]}{(\alpha_R-\beta_R)\alpha_R^{3/2}(4m^2-\alpha_R)^{1/2}}\times$$
$$\times\,\mathrm{arctg}\left(\frac{\alpha_R}{4m^2-\alpha_R}\right)^{1/2}-\frac{1}{\alpha_R} \tag{4.8}$$

The relation between the wave functions Φ_R and Φ_{NR} in the nonrelativistic limit with account of (4.3) takes the form

$$\Phi_R(x,\vec{p}_\perp)\rightarrow 2\pi^{1/2}m^{1/2}\,\Phi_{NR}(\vec{p})$$

Here, the relativistic wave function Φ_R is normalized by the condition (4.6) and the nonrelativistic wave function Φ_{NR} is normalized by the condition (4.4).

2. Break-up of The Relativistic Deuteron and The Verification of Scale Invariant Properties of Its Wave Function

Let us consider the interaction process of relativistic deuteron with target in which incoming deuteron breaks-up and a system of hadrons X_N is created. In the impulse approximation this process is described by two diagrams in which only one nucleon of the deuteron interacts with a target. Another nucleon, the so-called spectator-nucleon, doesn't interact with the target and observation of its distributions allows information on the character of internal motion of constituents in deuteron to be obtained. In the range of small transverse momenta this approximation seems very reasonable.

For the study of the dynamics of the relativistic deuteron, experiments with hydrogen targets are the most convenient, since in this case there are no effects associated with disintegration of the target, and the selection of the spectator nucleons makes it possible to obtain direct information about the deuteron wave function.

We consider now the process of interaction of the relativistic deuteron with hydrogen target in which a spectator-nucleon and the system X_N of hadrons are produced $D + p \to p_{sp}(n_{sp}) + X_N$. One of the simplest processes of this type is the process of direct break-up of the deuteron $Dp \to ppn$.

Assuming that the spectator-nucleon does not interact with the target we can obtain the following invariant distribution for spectator-nucleon [7,118,126]:

$$E_{sp}\frac{d\sigma}{d\vec{P}_{sp}} \sim \frac{\lambda^{1/2}(s_{NN},m^2,m^2)}{\lambda^{1/2}(s,m^2,m_D^2)}\sigma_{in}(s_{NN})\left|\frac{\Phi_R(x,\vec{p}_\perp)}{1-x}\right|^2 \qquad (4.9)$$

Here s is the usual Mandelstam variable for the deuteron-proton system,

s_{NN} is the similar variable for the subsystem consisting of the interacting nucleon of the deuteron and the target proton. Energy-momentum conservation leads to the following relation between these variables:

$$s_{NN}=s(1-X_{sp})+m^2-\frac{\vec{P}^2_{sp,\perp}+m^2}{X_{sp}} \tag{4.10}$$

$\sigma_{in}(s_{NN})$ is the total inelastic cross section of the nucleon-nucleon interaction in the given channel $NN \rightarrow X_N$. m is the nucleon mass, m_D is the deuteron mass. The flux factor is defined in the following way: $\lambda(x,y,z)=(x-y-z)^2-4yz$. The variable X_{sp} is defined as follows:

$$X_{sp}=\frac{(E_{sp}+P_{sp,z})}{(E_d+P_{d,z})+(E_p+P_{p,z})} \tag{4.11}$$

Here E_{sp}, E_d, E_p and $P_{sp,z}$, $P_{d,z}$, $P_{p,z}$ are the energies and the z-components of momenta of the spectator nucleon, colliding deuteron and proton, respectively. Note that the variable X_{sp} is scale-invariant and Lorenz-invariant under transformations of the reference frame along the collision axes (the z-axes).

The arguments of the wave function $\Phi_R(x,\vec{p}_\perp)$ are related to the variables X_{sp} and $\vec{P}_{sp,\perp}$ in the following way:

$$x=1-\left(1+\frac{E_p+P_{p,z}}{E_d+P_{d,z}}\right)X_{sp} \tag{4.12}$$

$\vec{p}_\perp=-\vec{P}_{sp,\perp}$ in the reference frame, in which $\vec{P}_{d,\perp}=0$

It follows from these expressions that the experimental observation of the spectator-nucleon distributions allows information on the internal motion of nucleons inside the relativistic deuteron to be obtained.

In the high energy limit the distribution of the spectator, summed over all possible hadron systems X_N, takes the form

$$\frac{1}{\sigma_{tot}(\infty)} \frac{d\sigma}{d\vec{P}_{sp}/E_{sp}}\Bigg|_{s\to\infty} \sim \frac{|\Phi_R(X_{sp},\vec{P}_{sp,\perp})|^2}{1-X_{sp}} \qquad (4.13)$$

which is very close to the predictions for inclusive distributions obtained in the framework of the hypothesis of limiting fragmentation [127] and those of the parton model [73-76] and of the automodelity principle for strong interactions [128]. Deviations from the automodel behaviour may occur becouse of a possible weak dependence of wave function parameters on the energy. So it seems very interesting the experimental study of processes with beams of deuterons of different energies. We shall discuss this point later on when comparing theoretical calculations with experimental data.

A characteristic feature of the spectator-nucleon distribution (4.9) in the reference frame, in which the target proton is at rest, is the prediction of a maximum at the value of variable X_{sp}

$$\tilde{X}_{sp} = \frac{1}{2\left(1+\frac{m}{E_d+P_{d,z}}\right)} \qquad (4.14)$$

which approaches its limiting value $\tilde{X}_{sp}=0{,}5$ as energy increases. As it will be seen from what follows the positions of the maxima of the experimental X_{sp}-distributions agree with the values predicted by (4.14).

To compare the relativistic parametrization (4.1) with the nonrelativistic Hulten wave function (4.2) we consider the momentum distribution of the spectator nucleons in the deuteron rest frame [129]. The momentum distribution of the spectators is related to the invariant differential cross section (4.9) in the following way

$$\frac{d\sigma}{dP_{sp}} = \frac{2\pi P_{sp}^2}{(P_{sp}^2 + m^2)^{1/2}} \int_{-1}^{1} d\cos\vartheta_{sp} \left(E_{sp} \frac{d\sigma}{d\vec{P}_{sp}} \right) \tag{4.15}$$

In Fig. 12 the results of theoretical calculations with the relativistic wave function (4.1) and with the nonrelativistic Hulten wave function (4.2) are compared with the experimental distribution of spectator neutrons in the process of direct break-up of the deuteron $D + p \rightarrow p + p + n_{sp}$ in the deuteron rest frame. The experimental data were obtained in the 100-cm hydrogen bubble chamber bombarded by deuterons with 3,3 GeV/c momentum and converted to the antilaboratory frame in which the momentum of the incident protons is equal to 1,66 GeV/c [130].

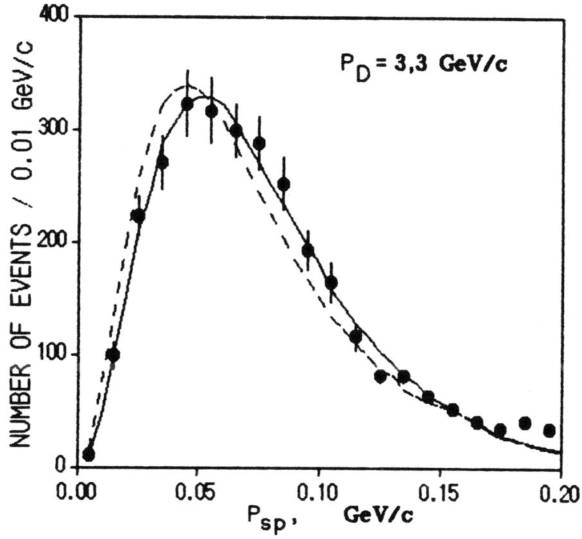

Fig. 12. Momentum distribution of the spectator neutrons in the deuteron rest frame in the reaction $Dp \rightarrow ppn_{sp}$. The solid and dashed curves are the results of calculation with the relativistic and the nonrelativistic wave functions, respectively.

Experimentally, the spectators were chosen as particles having the smallest momentum in the deuteron rest frame. The small admixture of other nucleons cannot affect significantly our consideration. Numerical values of the parameters α_R and β_R of the relativistic wave function of deuteron were calculated by Eq. (4.3) from the values α_{NR}=0,0456 GeV/c and β_{NR}=0,26 GeV/c [131] of the nonrelativistic Hulten wave function. The cross section of elastic nucleon-nucleon scattering at these energies is almost constant and equal to 24 mb [132]. The solid curve on Fig.12 corresponds to the theoretical calculation with the relativistic wave function (4.1) and the values of the parameters α_R=3,521 (GeV/c)^2 and β_R=3,390 (GeV/c)^2 (χ^2/N_p=27/20). The dashed curve corresponds to the calculations with the nonrelativistic wave function (4.2) and the parameters α_{NR}=0,0456 GeV/c and β_{NR}=0,26 GeV/c (χ^2/N_p=65/20). Comparison with experimental data shows that in the range of spectator momenta P_{sp} < 0,2 GeV/c the relativistic wave function gives somewhat better description of the experimental data than the nonrelativistic one.

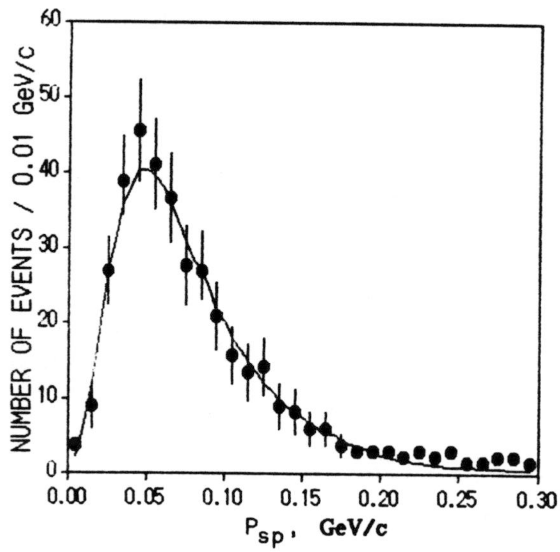

Fig. 13. Momentum distribution of the spectator protons in the deuteron rest frame in the reaction $Dp \rightarrow p(p\pi^-)p_{sp}$.

In Fig. 13 the momentum distribution of the spectator protons in the reaction $D + p \rightarrow p + (p\pi^-) + p_{sp}$ at the energy $\sqrt{s} = 52$ GeV is presented in the deuteron rest frame. The data were obtained in an experiment at the CERN ISR with the colliding deuteron-proton beams [30-31]. In Fig. 14 the momentum distribution of the spectator protons in the same reaction and at the same energy is presented in the frame of the colliding beams. The curves in Figs. 13 and 14 correspond to the calculations with the wave function (4.1). The cross section $\sigma_{in}(s_{NN})$ of neutron diffractive dissosiation $n + p \rightarrow (p\pi^-) + p$ at the considered energies was taken to be constant and equal to 185 mb [30]. The numerical values of the parameters α_R and β_R obtained by fitting the experimental data and the corresponding values of χ^2/N_p are given in Table 1. The agreement of the results of theoretical calculations with the experimental data confirms the validity of using the relativistic parametrization (4.1).

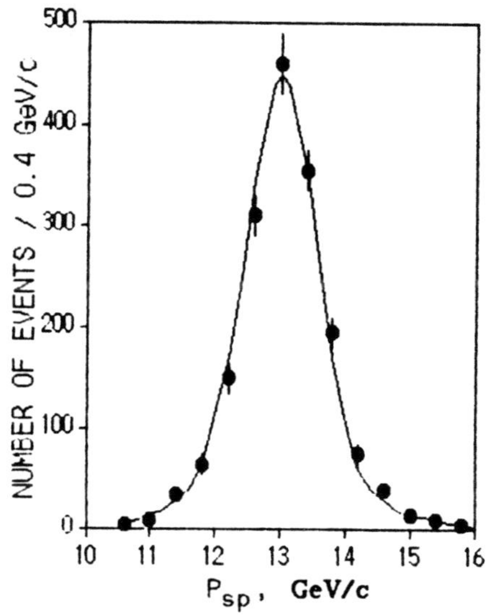

Fig. 14. Momentum distribution of the spectator protons in the frame of the colliding beams in the reaction $Dp \rightarrow p(p\pi^-)p_{sp}$.

Table 1. Parameters of the relativistic Hulthen wave function.

Reaction	α_R, (GeV/c)^2	β_R, (GeV/c)^2	χ^2/N_p
Calculation by Eq. (4.3)	3,521	3,390	
$D+p \rightarrow p+(p\pi^-)+p_{sp}$ $\sqrt{s}=52$ GeV Deuteron rest frame	3,522 ± 0,006	3,383 ± 0,033	11/32
$D+p \rightarrow p+(p\pi^-)+p_{sp}$ $\sqrt{s}=52$ GeV Center-of-mass frame	3,513 ± 0,005	3,478 ± 0,019	28/14
$D+Al \rightarrow n_{sp}+X_N$			
$P_D=3,46$ GeV/c	3,5156-fixed	3,5240 ± 0,0004	1,9
$P_D=4,46$ GeV/c	3,5156-fixed	3,5222 ± 0,0006	1,1
$P_D=7,66$ GeV/c	3,5156-fixed	3,5203 ± 0,0010	2,9
$P_D=10$ GeV/c	3,5156-fixed	3,5170 ± 0,0021	1,1
$D+p \rightarrow p+p+n_{sp}$ $P_D=3,3$ GeV/c $d\sigma/dX_{sp}$ in three intervals of $P_{sp,\perp}$	3,5156-fixed	3,4572 ± 0,0005	232/150
$D+p \rightarrow p+p+n_{sp}$ $P_D=3,3$ GeV/c $d\sigma/dP_{sp,\perp}$ in three intervals of X_{sp}	3,5156-fixed	3,4579 ± 0,0005	311/150

In order to verify the scaling properties of the relativistic wave functions, we have compared the theoretical calculations with the experimental distributions of the spectator nucleons at various momenta of incident deuterons [133]. The distributions of the spectator neutrons calculated using the wave function (4.1) were compared with the experimental distributions of neutrons from deuteron stripping, which were obtained in the following way. A beam of deuterons accelerated at the JINR Synchrophasotron to momentum P_D collided with an internal Alluminum target. Neutrons scattered from the target under the angles $\theta < 0{,}001$ rad were directed in the 100-cm bubble chamber. This set-up gives with a good accuracy the neutrons with zero transverse momentum $P_{sp,\perp}$. Momentum spectra of neutrons were measured in the reaction $np \to pp\pi^-$, which can be identified clearly in the bubble chamber. Events selected by this method were used for obtaining the momentum distributions of neutrons detected in the 100-cm bubble chamber. Detailed description of the experimental procedure can be found in [134-136].

Assuming that deuteron break-up process occurs on the separate nucleons of target nuclei one can suppose that distribution of spectator nucleons does not depend on the target. Independence of the spectator fragment distributions on the target for this type of processes seems to be the general property and allows one to use theoretical distribution (4.9) in these case.

In Figs. 15 a,b,c,d the experimental distributions

$$\left.\frac{d\sigma}{dX_{sp}\,d\Omega_{sp}}\right|_{\vec{P}_{sp,\perp}=0}$$

of spectator neutrons for four different values of the incident deuteron energy and the corresponding theoretical distributions normalized to the unity in the points of their maxima are presented. The relation between this

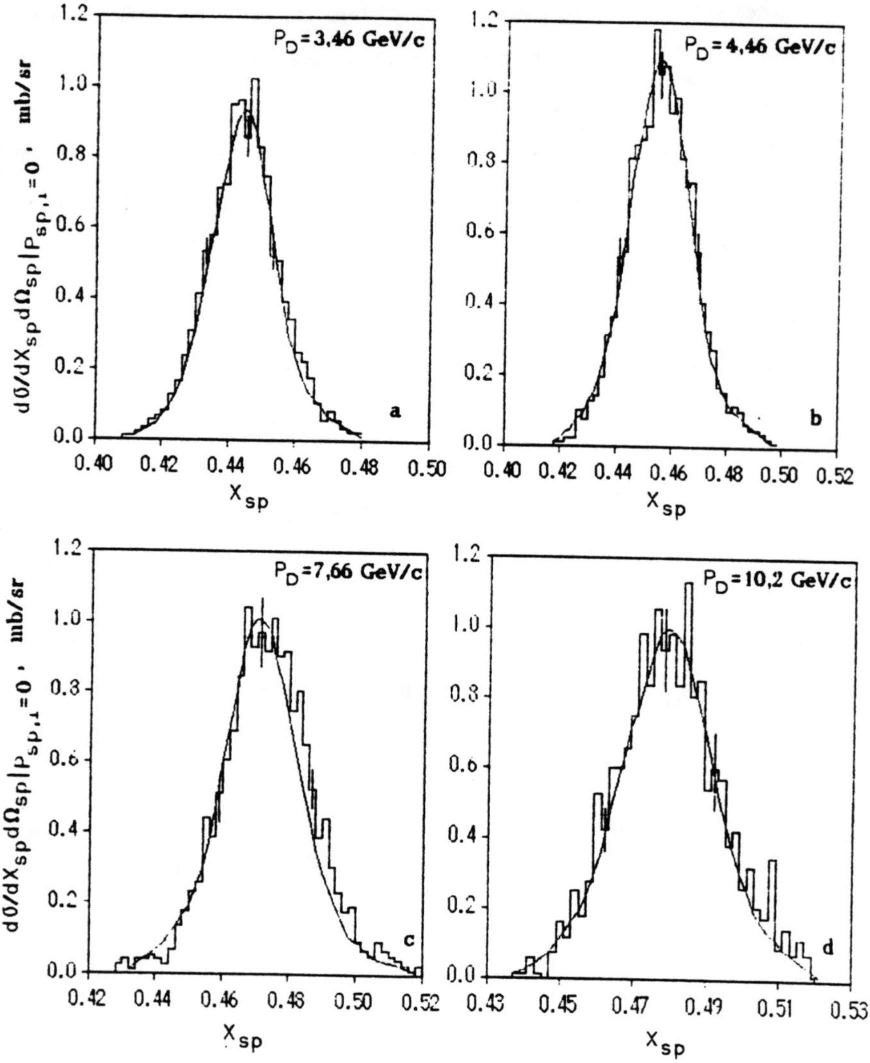

Fig. 15. The distributions $\frac{d\sigma}{dX_{sp}\,d\Omega_{sp}}\Big|_{P_{sp,\perp}=0}$ **of spectator neutrons at different momenta of the incident deuterons.**

distribution and the invariant distribution (4.9) at zero value of transverse momentum of the spectator has the form:

$$\frac{d\sigma}{dX_{sp}d\Omega_{sp}}\Bigg|_{\vec{P}_{sp,\perp}=0} = \frac{[(m+P_D+E_D)X_{sp}]^2-m^2}{4(m+P_D+E_D)^2X_{sp}^3}\left(E_{sp}\frac{d\sigma}{d\vec{P}_{sp}}\right)\Bigg|_{\vec{P}_{sp,\perp}=0}$$

An analysis performed shows that in the considered region of momenta of the incident deuteron the parameters of the relativistic wave function α_R and β_R depend weakly on the energy of the incident beam (see Table 1). This fact indicates that in wide range of energy the relativistic wave function of deuteron Φ_R contains no other dependence on the energy apart from the dependence on the variable x and allows to say that wave function with the scale-invariant parametrization of the "longitudinal motion" of constituents in terms of light front variables gives a good description of the relativistic deuteron.

The positions of maxima of the spectator neutron distributions in Figs. 15 a,b,c,d coinsides rather well with the values of $\tilde{X}_{sp}$ predicted by Eq. (4.14).

To study the transverse momentum distributions of the spectators [126] we have used the experimental data on the interaction of deuterons with momentum 3,3 GeV/c in hydrogen bubble chamber. From the experimental point of view, quite comprehensive data are available for the direct break-up channel $dp \rightarrow ppn$. In this case $\sigma_{in}(s_{NN})$ in Eq. (4.9) must be replaced by the total elastic cross section $\sigma_{el}(s_{NN})$ of the nucleon-nucleon interaction. We restricted ourselves to the case of the neutron-spectator, since there are much more experimental data on σ_{el}^{pp} than on σ_{el}^{np} (see, e.g., [132]). It should be noted that in the experiment with a beam of 3,3 GeV/c deuterons, s_{pp} varies in an interval, in which $\sigma_{el}(s_{pp})$ is almost constant and equal to 24 mb.

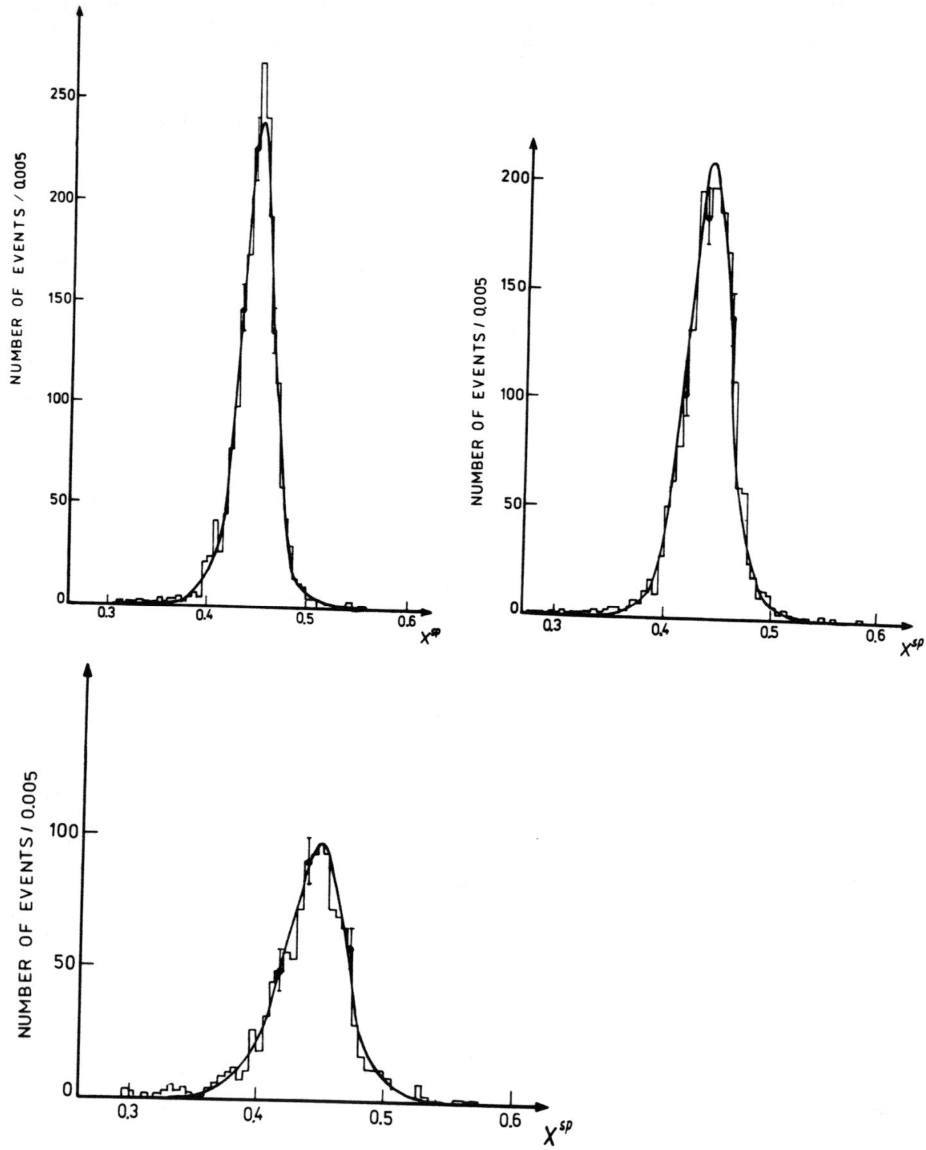

Fig. 16. The $d\sigma/dX_{sp}$ distributions of spectator neutron in the reaction $Dp \to ppn_{sp}$ in the intervals a) $0.01 < P_{sp,\perp} < 0.04$ GeV/c, b) $0.04 < P_{sp,\perp} < 0.07$ GeV/c, c) $0.07 < P_{sp,\perp} < 0.1$ GeV/c.

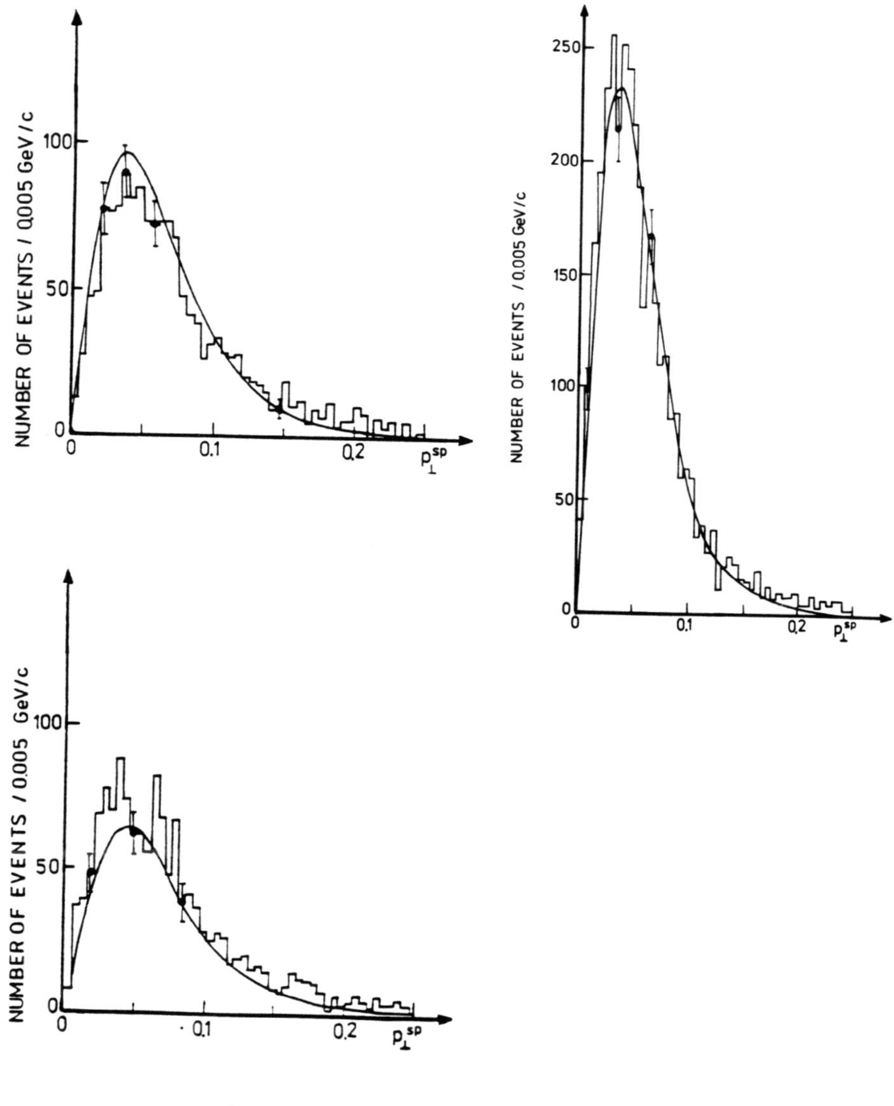

Fig. 17. **The** $d\sigma/dP_{sp,\perp}$ **distributions of spectator neutron in the reaction** $Dp \to ppn_{sp}$ **in the intervals a)** $0.40 < X_{sp} < 0.43$**, b)** $0.43 < X_{sp} < 0.46$**, c)** $0.46 < X_{sp} < 0.49$**.**

The experimental distributions $d\sigma/dX_{sp}$ and $d\sigma/dP_{sp,\perp}$ in the rest frame of the target proton were analysed. They are related to the invariant differential cross section as follows:

$$\frac{d\sigma}{dX_{sp}} = \int_{P_{sp,\perp\,min}}^{P_{sp,\perp\,max}} \frac{d\sigma}{dX_{sp}\,dP_{sp,\perp}}\,dP_{sp,\perp} \tag{4.16a}$$

$$\frac{d\sigma}{dP_{sp,\perp}} = \int_{X_{sp,min}}^{X_{sp,max}} \frac{d\sigma}{dX_{sp}\,dP_{sp,\perp}}\,dX_{sp} \tag{4.16b}$$

$$\frac{d\sigma}{dX_{sp}\,dP_{sp,\perp}} = 2\pi\left(\frac{P_{sp,\perp}}{X_{sp}}\right)\frac{d\sigma}{d\vec{P}_{sp}/E_{sp}} \tag{4.16c}$$

The experimental distributions $d\sigma/dX_{sp}$ $\left(d\sigma/dP_{sp,\perp}\right)$ integrated in three different intervals of $P_{sp,\perp}(X_{sp})$ were compared with the results of the theoretical scheme described above. The results of the analysis are given in Figs. 16 a,b,c and Figs 17 a,b,c. The theoretical curves in these Figures correspond to the values of parameters of the relativistic wave function (4.1) given in Table 1. The values of χ^2/N_p (where N_p is the number of experimental points) allow to consider the agreement of the model with the experimental data to be satisfactory. In accordance with the prediction (4.14), the maximum in the X_{sp} distribution is at $\tilde{X}_{sp} \approx 0.44$.

3. The Break Up of More Complicated Nuclei

It is interesting to attempt to find relativistic analogues of the wave functions of more complicated nuclei and to verify the universality of scale-invariant properties of these relativistic wave functions in the framework of light front formalism [25,137-139] by comparison of corresponding results with experimental data.

In this paragraph the process of knocking out the nucleon from the relativistic nucleus in the collision with the hydrogen target (proton) is considered. The relativistic nuclear wave functions having scale-invariant properties and turning in the nonrelativistic limit to the well-known Gaussian nuclear wave functions are introduced. The invariant differential cross section of the nucleon knock-out reaction from nucleus is expressed via the overlap integral of these wave functions. This overlap integral is calculated in an analitical form using the Gaussian parametrization of the light front wave functions. The results of theoretical calculations are compared with experimental data on the partial break-up of ^4He nuclei with momentum 8,56 GeV/c and ^3He nuclei with momentum 13,5 GeV/c in collisions with proton target. The data obtained on the 100-cm bubble chamber of the JINR [140,141] and the data on the process $^4He + p \rightarrow {}^3He + X$ at different value of momentum of incident nuclei obtained at SATURNE [142] have been used.

Let us consider the process of the knocking out one nucleon from the relativistic nucleus A on the hydrogen target. Assuming that only the knocked out nucleon interacts with the target and that the remaining (A-1) nucleons of the initial nucleus continue to exist in the form of a fragment nucleus (the so-called spectator fragment), one can calculate the distributions of these fragments.

The nucleus consisting of A nucleons with total 4-momentum P_A is described by means of the relativistic wave function $\Phi^{(A)}_{P_A}([x_i^{(A)}, \vec{p}_{i,\perp}])$ in which the "longitudinal motion" of the constituents is parametrized in terms of scale-invariant variables

$$x_i^{(A)} = \frac{p_0^{(i)} + p_3^{(i)}}{P_{A,0} + P_{A,3}}$$

where $p_\mu^{(i)}$ ($\mu = 0, 1, 2, 3$ is the Lorentz index) and $P_{A,\mu}$ are the individual 4-momentum of the i-th nucleon in the composite system and the total 4-momentum of the composite system as a whole. Here and in what follows the parameters (momenta, masses, etc.) of composite systems are denoted by upper-case letters, whereas the individual parameters of nucleons are denoted by lower-case letters. Square brackets in the argument of the wave function $\Phi_{P_A}^{(A)}$ denote the set of the corresponding variables $x_i^{(A)}$ and $\vec{p}_{i,\perp}$, which satisfy the following conditions:

$$\sum_{i=1}^{A} x_i^{(A)} = 1 \; ; \qquad 0 < x_i^{(A)} < 1 \; ; \qquad \sum_{i=1}^{A} \vec{p}_{i,\perp} = \vec{P}_{A,\perp} \tag{4.17}$$

The superscript of the variables $x_i^{(A)}$ means that this variable is defined in a system of particles whose number is equal to this index.

The distribution of the spectator fragments in the process of nucleon knock out from the nucleus in the laboratory frame (incoming nucleus A moves along the z-axis and the target proton is at rest) looks as follows

$$E_{sp} \frac{d\sigma}{d\vec{P}_{sp}} \sim \frac{\lambda^{1/2}(s_{NN}, m^2, m^2)}{\lambda^{1/2}(s, M_A^2, m^2)} \sigma_{NN}^{el}(s_{NN}) \left| \frac{I(X_{sp}, \vec{P}_{sp,\perp})}{1 - \alpha X_{sp}} \right|^2 \tag{4.18}$$

Here, s is the usual Mandelstam variable for the system consisting of the incident nucleus A and the target proton, s_{NN} is the analogous variable for subsystem consisting of the interacting nucleon of nucleus A and the target proton. Energy-momentum conservation leads to the following relation between these variables

$$s_{NN} = s(1 - X_{sp}) + M_{sp}^2 - \frac{\vec{P}_{sp,\perp}^{\,2} + M_{sp}^2}{X_{sp}} \tag{4.19}$$

$\sigma_{NN}^{el}(s_{NN})$ is the total elastic cross section for interaction of the nucleon from nucleus A with target. $\lambda(x,y,z)$ is the flux factor, which is defined as

$$\lambda(x,y,z) = (x - y - z)^2 - 4yz$$

m is the nucleon mass, M_A is the mass of the incident nucleus, M_{sp} is the mass of the spectator fragment. α is given by

$$\alpha = 1 + \frac{m}{E_A + P_{A,3}} \tag{4.20}$$

The variable X_{sp} is defined in the following way

$$X_{sp} = \frac{E_{sp} + P_{sp,3}}{m + E_A + P_{A,3}} \tag{4.21}$$

$P_{A,3}$, E_A and $P_{sp,3}$, E_{sp} are the z-components of the momenta and the energies of the incident nucleus A and of the spectator-fragment (A-1), respectively.

The overlap integral $I(X_{sp}, \vec{P}_{sp,\perp})$ of the relativistic wave functions of the incident nucleus and of the spectator fragment is given by:

$$I(X_{sp}, \vec{P}_{sp,\perp}) = \int_0^1 \prod_{i=1}^{A-1} \frac{dx_i^{(A-1)'}}{x_i^{(A-1)'}} \delta(1 - \sum_{i=1}^{A-1} x_i^{(A-1)'}) \times$$

$$\times \int \prod_{i=1}^{A-1} d\vec{p}_\perp^{\,(i)'} \delta^{(2)}(\vec{P}_{sp,\perp} - \sum_{i=1}^{A-1} \vec{p}_\perp^{\,(i)'}) \times \tag{4.22}$$

$$\times \Phi_f^{+(A-1)}([x_i^{(A-1)'}, \vec{p}_\perp^{\,(i)'} - x_i^{(A-1)'} \vec{P}_{sp,\perp}]) \, \Phi_{in}^{(A)}([x_i^{(A)}, \vec{p}_\perp^{\,(i)}])$$

The overlap integral is a direct analogue of the corresponding notion which appears in the nonrelativistic theory of nuclear reactions (see, e.g. Ref. [33] and the references therein).

The arguments of the wave function $\Phi_{in}^{(A)}$ of the incident nucleus are related to the integration variables and the observable quantities X_{sp} and $\vec{P}_{sp,\perp}$ in the following way:

$$\begin{aligned} & x_i^{(A)} = \alpha X_{sp} x_i^{(A-1)'} \; ; \qquad \vec{p}_\perp^{\,(i)} = \vec{p}_\perp^{\,(i)'} \; ; \qquad i = 1,2,\ldots,A-1 \\ & x_A^{(A)} = 1 - \alpha X_{sp} \; ; \qquad \vec{p}_\perp^{\,(A)} = -\vec{P}_{sp,\perp} \end{aligned} \tag{4.23}$$

Thus, observation of the spectator fragment makes it possible to obtain information about the character of the "longitudinal" and transverse momentum distributions of the nucleons in the incident nucleus A.

Note that the variable X_{sp} is defined in an arbitrary Lorentz frame, in which the nucleus A and the target proton collide along the z-axis as follows

$$X_{sp} = \frac{(E_{sp} + P_{sp,3})}{(E_N + P_{N,3}) + (E_A + P_{A,3})} \tag{4.24}$$

where P_3's and E's are the longitudinal momenta and the energies of the corresponding particles. As is easily seen, X_{sp} is a Lorentz-invariant and scale-invariant variable. In the proton rest frame it turns to the form given by Eq. (4.21).

In the case A=2, which corresponds to the process of the direct break-up of the deuteron, the overlap integral in Eq. (4.18) is replaced by the deuteron relativistic wave function.

If together with the knock-out of the nucleon from the nucleus the particle production takes place, the differential cross section of this process looks like the Eq. (4.18), but the cross section σ_{NN}^{el} of elastic scattering of the active nucleon from incident nucleus on the target nucleon is replaced by the cross section σ_{NN}^{tot} of particle production in the nucleon-nucleon scattering.

In the Eq. (4.22) the wave functions $\Phi_{in}^{(A)}$ and $\Phi_{f}^{(A-1)}$ of the incoming and outgoing nuclei are defined in the reference frame, in which their total transverse momenta are zero. They are related to the wave functions with arbitrary total 4-momentum P as follows [52]

$$\Phi_{P}^{(A)}([x_i^{(A)}, \vec{p}_{\perp}^{(i)}]) = \Phi_{P_{\perp}=0}^{(A)}([x_i^{(A)}, \vec{p}_{\perp}^{(i)} - x_i^{(A)}\vec{P}_{\perp}]) \qquad (4.25)$$

As an example, we consider the following simplest parametrization of the relativistic wave functions of the incident and fragment nuclei:

$$\Phi^{(A)}([x_i^{(A)}, \vec{p}_{\perp}^{(i)}]) = C_A \exp\left(-a_A^R \sum_{i=1}^{A} \frac{\vec{p}_{\perp}^{(i)2} + m_i^2}{x_i^{(A)}}\right) \qquad (4.26)$$

and similarly for $\Phi^{(A-1)}$, with the substitution $A \to (A-1)$.

In Eq. (4.26) a_A^R is an adjustable parameter, C_A is a normalization factor. If the scale-invariant parametrization of "longitudinal motion" of the constituents put into the wave function $\Phi^{(A)}([x_i^{(A)}, \vec{p}_{\perp}^{(i)}])$ is really valid, then the parameters a_A^R would be the same at different energies of incoming nuclei.

Since we do not distinguish protons from neutrons in our consideration, the wave function $\Phi^{(A)}([x_i^{(A)}, \vec{p}_{\perp}^{(i)}])$ is a symmetric function of its arguments. Solving the problem of a conditional extremum under the conditions (4.17), we find that the wave function (4.26) has a maximum at zero values of the transverse momenta of the constituent nucleons and at values of the variables $x_i^{(A)}$ equal to

$$\tilde{x}_i^{(A)} = \frac{m_i}{\sum_{i=1}^{A} m_i} = \frac{1}{A} \tag{4.27}$$

Taking into account the relation between the variables $x_i^{(A)}$ and X_{sp} (see Eqs. (4.23)), we find that the X_{sp} distributions of the spectator fragments must have a maximum at

$$\tilde{X}_{sp} = \frac{A-1}{A\left(1+\frac{m}{E_A+P_{A,3}}\right)} \tag{4.28}$$

Note that properties such as the scale invariance of the relativistic wave functions and the position of the maximum in the X_{sp} distribution of the spectator fragments, do not depend on the concrete form of (4.26) of the wave functions $\Phi^{(A)}([\, x_i^{(A)}, \vec{p}_\perp^{(i)}])$ and remain valid for their arbitrary parametrization.

In order to normalize the relativistic wave functions correctly, one has to know in general the form of all the interactions inside the relativistic system. Assuming, however, that the total interaction kernel does not depend on the total 4-momentum of the composite system, one gets the following normalization condition:

$$\int_0^1 \prod_{i=1}^{A} \frac{dx_i^{(A)}}{x_i^{(A)}} \delta(1-\sum_{i=1}^{A} x_i^{(A)}) \int \prod_{i=1}^{A} d\vec{p}_\perp^{(i)} \delta^{(2)}(\vec{P}_{A,\perp} - \sum_{i=1}^{A} \vec{p}_\perp^{(i)}) \times$$
$$\times |\Phi_{P_A}^{(A)}([x_i^{(A)}, \vec{p}_\perp^{(i)}])|^2 = 2(4\pi)^{A-1} \tag{4.29}$$

Substituting the wave function $\Phi^{(A)}([x_i^{(A)}, \vec{p}_\perp^{(i)}])$ in the form of (4.26) into the normalization condition (4.29), we obtain the following approximate expression for the normalization factor C_A:

$$C_A \simeq \sqrt{2}\,(4\pi)^{\frac{A-1}{2}} \left(\sum_{i=1}^{A} m_i\right)^{\frac{3(A-1)+1}{4}} \left(\prod_{i=1}^{A} m_i\right)^{-\frac{1}{4}} \times$$
$$\times \left(\frac{2a_A^R}{\pi}\right)^{\frac{3(A-1)}{4}} \exp\left[a_A^R\left(\sum_{i=1}^{A} m_i\right)^2\right] \tag{4.30}$$

One of the guiding points in the choice of relativistic wave functions is their correct nonrelativistic limit. The nonrelativistic wave function $\Phi_{NR}^{(A)}([\vec{p}^{(i)}])$, the relativistic analogue of which is given by Eq. (4.26) has the Gaussian form

$$\Phi_{NR}^{(A)}([\vec{p}^{(i)}]) = \left(\frac{4a_A^{NR}}{\pi}\right)^{\frac{3(A-1)}{4}} \exp\left(-a_A^{NR}\sum_{i=1}^{A}\vec{p}^{(i)2}\right) \tag{4.31}$$

and is normalized by the condition

$$\int \prod_{i=1}^{A} d\vec{p}^{(i)}\, \delta^{(3)}\left(\sum_{i=1}^{A}\vec{p}^{(i)}\right) \left|\Phi_{NR}^{(A)}([\vec{p}^{(i)}])\right|^2 = 1 \tag{4.32}$$

Let us expand the combination $(\vec{p}_\perp^{(i)2} + m_i^2)/x_i^{(A)}$ in the power series of the parameters $p_3^{(i)}/m_i$, $p_\perp^{(i)}/m_i$, which are small in the nonrelativistic limit. Restricting ourselves to the quadratic terms and writting the result in the nucleus rest frame $\vec{P}_A = 0$, we get

$$\frac{\vec{p}_\perp^{(i)2} + m_i^2}{x_i^{(A)}} \simeq (m_i - p_3^{(i)})\left(\sum_{i=1}^{A} m_i\right) + \frac{\vec{p}^{(i)2}}{2m_i}\left(\sum_{i=1}^{A} m_i\right) +$$
$$+ m_i\left(\sum_{i=1}^{A}\frac{\vec{p}^{(i)2}}{2m_i}\right) \tag{4.33}$$

Substituting this expression into the wave function (4.26) we obtain in the nonrelativistic limit (without taking into account the normalization factor)

$$\Phi^{(A)}([x_i^{(A)}, \vec{p}_\perp^{(i)}]) \sim \exp\left(-a_A^R \sum_{i=1}^{A} \frac{\vec{p}_\perp^{(i)2} + m_i^2}{x_i^{(A)}}\right) \rightarrow$$

$$\rightarrow \exp\left[-a_A^R\left(\sum_{i=1}^{A} m_i\right)^2\right] \exp\left[-a_A^R\left(\sum_{i=1}^{A} m_i\right)\left(\sum_{i=1}^{A} \frac{\vec{p}^{(i)2}}{m_i}\right)\right] \tag{4.34}$$

The condition of the correct nonrelativistic limit gives the following relation between the parameters of the relativistic and nonrelativistic wave functions

$$a_A^R = \frac{m_i}{\sum_{i=1}^{A} m_i} = \frac{1}{A} a_A^{NR} \tag{4.35}$$

For the normalized wave functions, the nonrelativistic limit looks as follows

$$\Phi_R^{(A)}([x_i^{(A)}, \vec{p}_\perp^{(i)}]) \rightarrow \sqrt{2}\,(2\pi^2 m^2)^{\frac{A-1}{4}} A^{\frac{1}{4}}\, \Phi_{NR}^{(A)}([\vec{p}^{(i)}]) \tag{4.36}$$

Here $\Phi_R^{(A)}$ is the relativistic wave function (4.26) normalized by the condition (4.29) and $\Phi_{NR}^{(A)}$ is the nonrelativistic wave function (4.31) normalized by the condition (4.32). Masses of all nucleons are assumed to be equal $m_i = m$.

Substituting now the wave functions $\Phi^{(A)}$ and $\Phi^{(A-1)}$ into the overlap integral (4.22), taking into account the relations (4.23) and integrating over the transverse momenta, we obtain

$$I(X_{sp}, \vec{P}_{sp,\perp}) = C_A C_{A-1}\left(\frac{\pi}{a_{A-1}^R + a_A^R/(\alpha X_{sp})}\right)^{A-2} \times$$

$$\times \exp\left(-\frac{a_A^R m_A^2}{1-\alpha X_{sp}}\right) \exp\left[-\frac{a_A^R \vec{P}_{sp,\perp}^2}{\alpha X_{sp}(1-\alpha X_{sp})}\right] J(X_{sp}) \tag{4.37}$$

where

$$J(X_{sp}) = \int_0^1 \prod_{i=1}^{A-1} dx_i^{(A-1)} \delta\left(1 - \sum_{i=1}^{A-1} x_i^{(A-1)}\right) \times$$
$$\times \exp\left[-\left(a_{A-1}^R + \frac{a_A^R}{\alpha X_{sp}}\right) \sum_{i=1}^{A-1} \frac{m_i^2}{x_i^{(A-1)}}\right] \qquad (4.38)$$

Thus, the $\vec{P}_{sp,\perp}$-distribution is obtained in the analytic form. The integral over the variables $x_i^{(A-1)}$ can be calculated approximately by means of the multidimensional method of steepest descent [143] in the form of an asymptotic expansion in inverse powers of the large parameter $a_{A-1}^R + a_A^R/\alpha X_{sp}$. The leading term of this expansion has the form:

$$J_0(X_{sp}) \simeq \left(\prod_{i=1}^{A-1} m_i\right)^{1/4} \left(\sum_{i=1}^{A-1} m_i\right)^{-\frac{3(A-2)+1}{2}} \left(\frac{\pi}{a_{A-1}^R + a_A^R/(\alpha X_{sp})}\right)^{\frac{A-2}{2}} \times$$
$$\times \exp\left[-\left(a_{A-1}^R + \frac{a_A^R}{\alpha X_{sp}}\right)\left(\sum_{i=1}^{A-1} m_i^2\right)^2\right] \qquad (4.39)$$

One can estimate the first order correction to the value of integral (4.38). With account of this correction the integral over the variables $x_i^{(A-1)}$ takes the following form

$$J(X_{sp}) = J_0(X_{sp}) K(X_{sp}) \qquad (4.40)$$

where

$$K(X_{sp}) = 1 - \frac{3}{16}\left(\sum_{i=1}^{A-1} m_i\right)^{-2} \left(a_{A-1}^R + \frac{a_A^R}{\alpha X_{sp}}\right)^{-1} \times$$
$$\times \left[9\left(\sum_{i=1}^{A-1} m_i\right)\left(\sum_{i=1}^{A-1} \frac{1}{m_i}\right) - 3(A-1)^2 - 14(A-1) + 8\right] \qquad (4.41)$$

Substituting into Eq. (4.37) the expressions for the normalization factors C_A and C_{A-1} from (4.30) and the expression for the integral $J(X_{sp})$, we obtain finally for the overlap integral

$$I(X_{sp},\vec{P}_{sp,\perp})=2(4\pi)^{\frac{2A-3}{2}}\left(\frac{2a_A^R}{\pi}\right)^{\frac{3(A-1)}{4}}\left(\frac{2a_{A-1}^R}{\pi}\right)^{\frac{3(A-2)}{4}}\times$$

$$\times\left(\sum_{i=1}^{A}m_i\right)^{\frac{3(A-1)+1}{4}}\left(\sum_{i=1}^{A-1}m_i\right)^{-\frac{3(A-2)+1}{4}}\left(\prod_{i=1}^{A}m_i\right)^{-\frac{1}{4}}\left(\prod_{i=1}^{A-1}m_i\right)^{\frac{1}{4}}\times$$

$$\times\left(\frac{\pi}{a_{A-1}^R+a_A^R/(\alpha X_{sp})}\right)^{\frac{3(A-2)}{2}}\exp\left[-\frac{a_A^R\vec{P}_{sp,\perp}^{\,2}}{\alpha X_{sp}(1-\alpha X_{sp})}\right]\times \tag{4.42}$$

$$\times\exp\left\{-a_A^R\frac{\left[\left(\sum_{i=1}^{A}m_i\right)\alpha X_{sp}-\left(\sum_{i=1}^{A-1}m_i\right)\right]^2}{\alpha X_{sp}(1-\alpha X_{sp})}\right\}K(X_{sp})$$

It can be seen from the overlap integral (4.42) that the distribution with respect to the transverse momentum $\vec{P}_{sp,\perp}$ of the spectator fragment must have a Gaussian form and X_{sp}-distribution will have a maximum at the point predicted by Eq. (4.28).

Now the Eq. (4.18) for the differential cross section of the nucleon knock-out from the relativistic nucleus with the overlap integral (4.42) can be used for a comparison with experimental data.

4. The Scaling Properties of the Relativistic Wave Functions of ^4He and ^3He Nuclei

To verify the scaling properties of the relativistic wave functions and to extract the information about the values of their parameters, we used the experimental data on $^4He\,p$ interaction at 8,56 GeV/c and 13,5 GeV/c momenta of incident ^4He nucleus and $^3He\,p$ interaction at 13,5 GeV/c momentum of incident ^3He nucleus. These data were obtained in the 100-cm hydrogen bubble chamber [140,141] of the JINR at Dubna. We also used the data on the momentum distribution of the ^3He nuclei emitted at angle $0^{\circ}65'$ in the $^4He\,p \to {}^3He\,X$ reaction with an incident ^4He nucleus having momentum 6,85 GeV/c [142].

The experimental $d\sigma/dX_{sp}$ and $d\sigma/dP_{sp,\perp}$ distributions of the ^3H and ^3He spectator fragments in the $^4He\,p \to {}^3H\,p\,p$ and $^4He\,p \to {}^3He\,p\,n$ reactions were analysed in the rest frame of the target proton. (The spectator nuclei ^3H and ^3He were defined as the fragments having the smallest momentum among the reaction products in the ^4He rest frame). They are related to the invariant differential cross section (4.18) as follows

$$\frac{d\sigma}{dX_{sp}} = \int_0^{P_{sp,\perp\,max}} \frac{d\sigma}{dX_{sp}\,dP_{sp,\perp}}\,dP_{sp,\perp} \qquad (4.43\,a)$$

$$\frac{d\sigma}{dP_{sp,\perp}} = \int_{X_{sp\,min}}^{X_{sp\,max}} \frac{d\sigma}{dX_{sp}\,dP_{sp,\perp}}\,dX_{sp} \qquad (4.43\,b)$$

$$\frac{d\sigma}{dX_{sp}\,dP_{sp,\perp}} = 2\pi\left(\frac{P_{sp,\perp}}{X_{sp}}\right)\frac{d\sigma}{d\vec{P}_{sp}/E_{sp}} \qquad (4.43\,c)$$

The limits of integration $X_{sp_{min}}$ and $X_{sp_{max}}$ were taken from the corresponding experimental X_{sp}-distributions of the spectator fragments. The upper limit $P_{sp,\perp_{max}}$ is the kinematical limit determined by the condition of the positivity of the factor $\lambda(s_{NN}, m^2, m^2)$ in Eq. (4.18):

$$\left(P_{sp,\perp_{max}}\right)^2 = (sX_{sp} - M_{sp}^2)^2(1 - X_{sp}) - 4m^2X_{sp} \tag{4.44}$$

The total elastic cross section σ_{NN}^{el} can be assumed to be effectively constant in the considered range of energies and approximately equal to 24 mb [132].

The overlap integral (4.42) depends weakly on the values of the parameter of the spectato nucleus. Therefore, when the parameters of the relativistic wave functions of the final nuclei ^3H and ^3He are determined by fitting the experimental data, there are large errors. For this reason, the fitting was done for a fixed value of the parameter of the wave function of the three-nucleon nucleus.

For determination of the parameter a_3^R we used the relation (4.35) between the parameters of relativistic and nonrelativistic wave functions. It should be noted that the parameters of the nonrelativistic Gaussian wave functions of ^4He, ^3He and ^3H nuclei are not determined well enough and we do not have reliable information about them. The values of the parameters a_4^{NR} and a_3^{NR}, given in the literature, vary in rather wide range (see, e.g., Refs. [144-147]). The greatest number of data are available for the parameters of the nonrelativistic ^4He wave function, but the values of the parameter a_4^{NR} of the Gaussian parametrization (4.31) vary in the range $a_4^{NR} = 20 - 28\ (GeV/c)^{-2}$.

The experimental and theoretical $P_{sp,\perp}$ and X_{sp}-distributions of the spectator nucleus ^3H in the $^4He\,p \rightarrow {}^3H\,p\,p$ reaction at 8,56 GeV/c momentum of the incident ^{4}He nucleus are given in Fig. 18. The same distributions of the ^{3}He spectator fragments in the ^{4}He p → ^{3}He p n reaction at the same momentum are presented in Fig. 19. The $P_{sp,\perp}$ and X_{sp} distributions of ^{3}H fragments in ^{4}He p → ^{3}H p p reaction at 13,5 GeV/c momentum of incident ^{4}He are given in Fig. 20 and momentum spectrum of ^{3}He fragments scattered on 0°65 in the laboratory frame in the ^{4}He p → ^{3}He X reaction at 6,85 GeV/c momentum of incident ^{4}He nucleus is shown in Fig. 21. The solid curves in Figs. 18 - 21 correspond to the theoretical calculations with values of the parameters obtained by fitting the experimental data and presented in Table 2. As can be seen from this Table the values of parameter a_4^R of the relativistic wave function obtained by fitting the data at different momenta of the incident nucleus are close to each other and agree satisfactorily with the values $a_4^R \approx 5 - 7$ (GeV/c)$^{-2}$, predicted by Eq. (4.35).

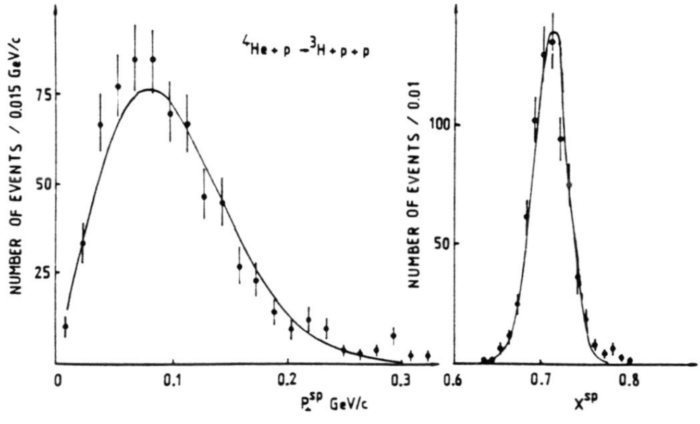

Fig. 18. The $P_{sp,\perp}$ and X_{sp} distributions of the spectator fragment ^{3}H in the ^{4}He p → ^{3}H p p reaction at ^{4}He momentum 8.56 GeV/c.

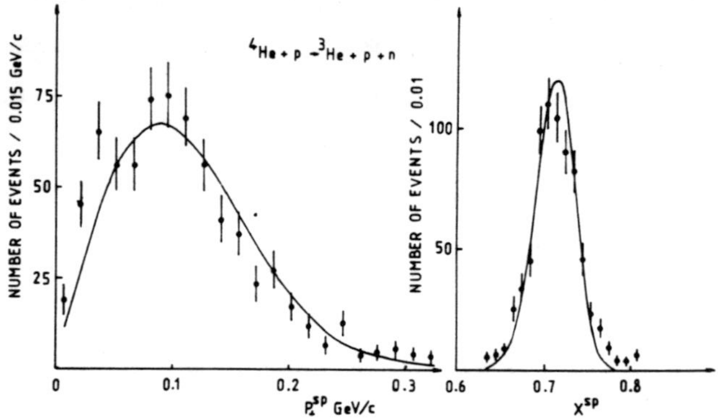

Fig. 19. The $P_{sp,\perp}$ and X_{sp} distributions of the spectator fragment ^{3}He in the ^{4}Hep → ^{3}Hepp reaction at ^{4}He momentum 8.56 GeV/c.

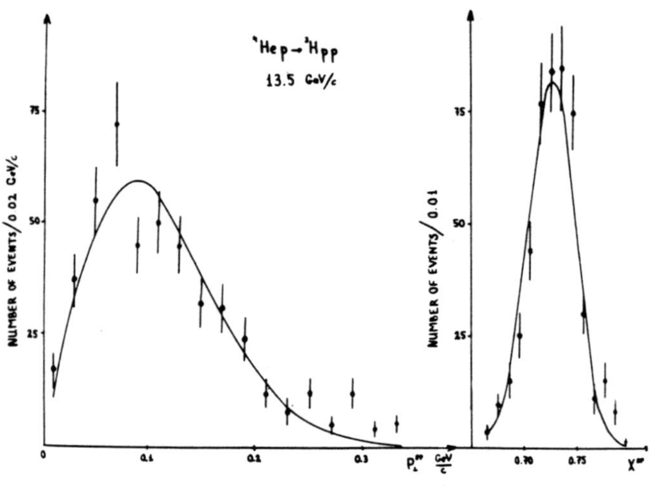

Fig. 20. The $P_{sp,\perp}$ and X_{sp} distributions of the spectator fragment ^{3}H in the reaction ^{4}Hep → ^{3}Hpp at ^{4}He momentum 13.5 GeV/c.

The positions of the maxima in the X_{sp} distributions of the spectator fragments agree well with the values of $\tilde{X}_{sp}$ predicted by Eq. (4.28).

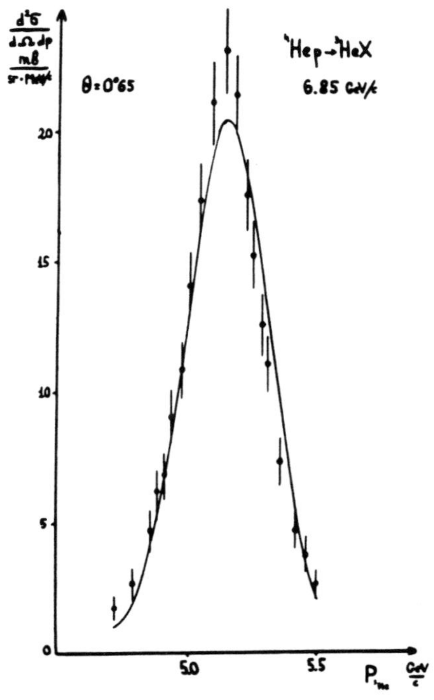

Fig. 21. Momentum spectrum of ^{3}He fragments emitted at angle 0°65' in the ^{4}Hep → ^{3}HeX reaction at ^{4}He momentum 6.85 GeV/c.

The approximate energy-independence of the values of the parameters of the relativistic wave function in the considered region of energies of the incident ^{4}He nucleus allows to assume that in the wave function (4.26) there is no other dependence on energy apart from the dependence on the scale-invariant variables $x_i^{(A)}$.

Table 2. The parameters of relativistic wave functions of three- and four-nucleon nuclei

Reaction, momentum	a_3^R, $(GeV/c)^{-2}$	a_4^R, $(GeV/c)^{-2}$	χ^2/N_p	
			$d\sigma/dX_{sp}$	$d\sigma/dP_{sp,\perp}$
$^4He\,p \to {}^3H\,p\,p$ P_{He}=8,56 GeV/c	8 - fixed	$7{,}39 \pm 0{,}26$	31,12/18	27,65/22
$^4He\,p \to {}^3He\,p\,n$ P_{He}=8,56 GeV/c	8 - fixed	$5{,}86 \pm 0{,}21$	54,63/18	43,15/22
$^4He\,p \to {}^3H\,p\,p$ P_{He}=13,5 GeV/c	8 - fixed	$6{,}23 \pm 0{,}25$	29,01/14	37,14/17
$^4He\,p \to {}^3He\,X$ P_{He}=6,85 GeV/c	8 - fixed	$7{,}21 \pm 0{,}34$	34,93/20	

For the experimental determination of the parameter a_3^R of the relativistic wave function of the three-nucleon nucleus, we consider the $^3He\,p \to D\,p\,p$ reaction. In this case, the overlap integral has the following form

$$I(X_{sp}, \vec{P}_{sp,\perp}) = 4\pi C_D 3^{7/4} m (a_3^R)^{3/2} \times$$

$$\times \exp\left[-\frac{a_3^R \vec{P}^2_{sp,\perp}}{\alpha X_{sp}(1-\alpha X_{sp})}\right] \exp\left[-a_3^R \frac{m^2(3\alpha X_{sp}-2)^2}{\alpha X_{sp}(1-\alpha X_{sp})}\right] \times \tag{4.45}$$

$$\times \sum_{i=1}^{5} A_i^R \exp(-4m^2\alpha_i^R)\left(\alpha_i^R + \frac{a_3^R}{\alpha X_{sp}}\right)^{-3/2}$$

Here C_D is a normalization factor

$$C_D = 2^{5/2}\pi^{-1/4} m^{1/2}\left\{\sum_{i=1}^{5}\sum_{j=1}^{5}\frac{A_i^R A_j^R}{(\alpha_i^R+\alpha_j^R)^{3/2}}\times\right.$$

$$\left.\times\exp\left[-4m^2(\alpha_i^R+\alpha_j^R)\right]\right\}^{-1/2}$$

and A_i^R and α_i^R are the parameters of the relativistic wave function of deuteron:

$$\Phi_D^R(x,\vec{p}_\perp) = C_D\sum_{i=1}^{5} A_i^R \exp\left[-\alpha_i^R\frac{\vec{p}_\perp^{\,2}+m^2}{x(1-x)}\right] \tag{4.46}$$

which we chose in the form of the relativistic analog of the Gartenhaus-Moravcsik wave function

$$\Phi_D^{NR}(\vec{p}) = C_D^{NR}\sum_{i=1}^{5} A_i^{NR}\exp\left(-\alpha_i^{NR}\vec{p}^{\,2}\right) \tag{4.47}$$

The parameters A_i^R and α_i^R are related to the parameters of the nonrelativistic wave function (4.47) in the following way

$$A_i^R = A_i^{NR}\exp(m^2\alpha_i^{NR}),\qquad \alpha_i^R = \frac{1}{4}\alpha_i^{NR} \tag{4.48}$$

When comparing the theoretical calculation with the experimental data on the distribution of the spectator deuterons in the reaction $^3He\,p \rightarrow D\,p\,p$ the parameters of the deuteron wave function (4.46) were fixed by the values obtained in accordance with the expression (4.48) from the values of the parameters of the nonrelativistic wave function (4.47) given in Ref. [148].

In Fig. 22 the X_{sp}-distribution of spectator deuterons in the $^3He\,p \rightarrow D\,p\,p$ reaction at 13,5 GeV/c momentum of incident ^3He is presented.

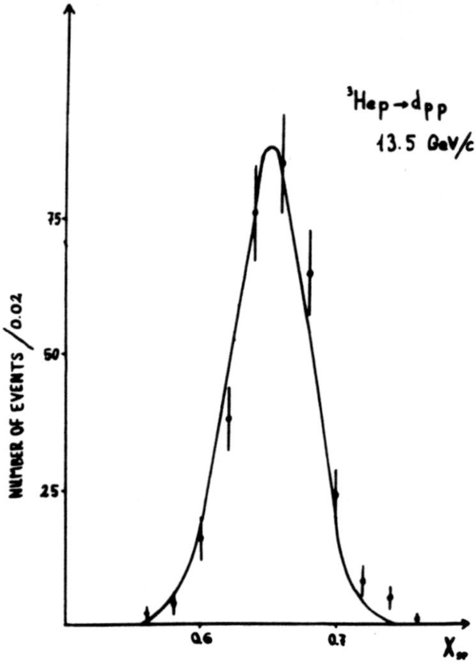

Fig. 22. The X_{sp} distribution of spectator deuterons in the reaction ^{3}Hep → Dpp at ^{3}He momentum 13.5 GeV/c.

The theoretical curve corresponds to the calculation with the overlap integral (4.45) with the value of parameter $a_3^R = 8{,}28 \pm 0{,}71\ (\mathrm{GeV}/c)^{-2}$, obtained by fitting the experimental data. This value of the parameter a_3^R agrees well with the value $a_3^R = 8\ (\mathrm{GeV}/c)^{-2}$, predicted by Eq. (4.35) and used by us in previos calculations.

It is interesting to compare the theoretical calculations on the distributions of the spectator deuterons with experimental data at different energies of incident ^{3}He nucleus in order to check more precisely the scaling properties of the relativistic wave function of ^3He nucleus. It is also of interest to consider wave functions of a more complicated form. However, our simplest parametrizations of the relativistic nuclear wave functions make it possible to establish a number of features (such as the scale-invariant character of "longitudinal" motion of nucleons inside the nucleus) that do not depend on their specific parametrization.

Chapter V

DEEP INELASTIC LEPTON-NUCLEUS SCATTERING WITH THE FRAGMENT NUCLEUS IN THE FINAL STATE

In this chapter we present some results on lepton-nucleus interaction, obtained using the many-body formalism in the light front variables. We'll discuss deep inelastic lepton-nucleus processes when the final lepton and the spectator fragment of the initial nucleus are detected in coincidence [149,150].

1. Deep Inelastic Scattering of Charged Lepton on Nucleus. The Process $lA \to l'(A-1)X$.

We consider the process of the deep inelastic scattering of the electron (muon) with 4-momentum k by a nucleus A with 4-momentum P_A when the final lepton with 4-momentum k' and the spectator fragment $(A-1)$ of the nucleus A with 4-momentum P^{sp}_{A-1} are detected in coincidence in the final state. The kinematic inva- riants of the process $lA \to l'(A-1)X$ are defined in the following manner:

$$q^2 = (k-k')^2, \qquad \nu_A = P_A q, \qquad \nu_{sp} = P^{sp}_{A-1} q, \qquad x = P_A P^{sp}_{A-1}, \quad s_{lA} = (k+P_A)^2 \tag{5.1}$$

The elements of the phase space of the particles detected in the final state are expressed in terms of these invariants as

$$\frac{d\vec{k}'}{E_{k'}} = 2\pi\,\lambda^{-1/2}(s_{lA}, m_l^2, m_A^2)\,dq^2 d\nu_A$$
$$\frac{d\vec{P}^{sp}_{A-1}}{E^{sp}_{A-1}} = (\nu_A^2 - m_A^2 q^2)^{-1/2}\,d\nu_{sp}\,dx\,d\varphi_{sp} \qquad (5.2)$$

m_A is the nucleus mass, m_l is the lepton mass, $\lambda(x,y,z)=(x-y-z)^2-4yz$. Note that before proceeding to a more precise definition the process is considered in an arbitrary Lorentz frame in which the virtual photon and the initial nucleus A move along the z-axis, i.e.

$$q=(q_0,0,0,q_z), \qquad P_A=(E_A,0,0,P_{A,z}),$$
$$P^{sp}_{A-1}=(E^{sp}_{A-1}, P^{sp}_{A-1,\perp}\sin\varphi_{sp}, P^{sp}_{A-1,\perp}\cos\varphi_{sp}, P^{sp}_{A-1,z}).$$

In the lowest order in the electromagnetic interaction the matrix element of the process under consideration is of the form:

$$T(lA\to l'(A-1)X)=\frac{4\pi\alpha}{q^2}\, j_\mu \langle P^{sp}_{A-1}, X | J_\mu(0) | P_A\rangle \qquad (5.3)$$

α is the fine structure constant, $j_\mu=\bar{u}(k')\gamma_\mu u(k)$ is the leptonic electromagnetic current, J_μ is the nucleus electromagnetic current.

The differential cross section of the deep inelastic lepton-nucleus scattering in the one-photon exchange approximation looks as follows:

$$\frac{d\sigma(lA\to l'(A-1)X)}{dq^2 d\nu_A d\nu_{sp} dx\, d\varphi_{sp}} = \frac{(\nu_A^2-m_A^2q^2)^{-1/2}}{8(2\pi)^5\,\lambda(s_{lA}, m_l^2, m_A^2)}\left(\frac{4\pi\alpha}{q^2}\right)^2 l_{\mu\nu}\,\hat{W}^{lA}_{\mu\nu}. \qquad (5.4)$$

The tensor $l_{\mu\nu}$ defining the leptonic part of the process is the product of electromagnetic currents of the leptons averaged over the spin states of the initial lepton and summed over the spin states of the final lepton:

$$l_{\mu\nu} = \sum_{spin} j_\mu^* j_\nu = [2k_\mu k'_\nu + 2k_\nu k'_\mu - 2g_{\mu\nu}(kk' - m_l^2)] \tag{5.5}$$

All the information about the strongly interacting part of the process considered is contained in the tensor $\hat{W}^{IA}_{\mu\nu}$, which is defined in the following manner:

$$\hat{W}^{IA}_{\mu\nu}(P_A, q, P^{sp}_{A-1}) = \sum_{spin} \sum_X (2\pi)^4 \delta^{(4)}(P_A + q - P^{sp}_{A-1} - p_X) \times \\ \times \langle P_A | J_\mu(0) | P^{sp}_{A-1}, X \rangle \langle P^{sp}_{A-1}, X | J_\nu(0) | P_A \rangle \tag{5.6}$$

Here the sum over X means summation over all possible states with a spectator fragment in the final state and summation over the spin states and integration over the phase spaces of the undetected final hadrons for each of such states.

Conservation of the electromagnetic current leads to the following condition of gauge invariance for the tensor $\hat{W}^{IA}_{\mu\nu}$:

$$q_\mu \hat{W}^{IA}_{\mu\nu} = \hat{W}^{IA}_{\mu\nu} q_\nu = 0 . \tag{5.7}$$

Taking into account the symmetry of $l_{\mu\nu}$ under the $\mu \to \nu$ exchange (see Eq. (5.5)), it can be seen easily that a contribution to the cross section is given only by the symmetric part of $\hat{W}^{IA}_{\mu\nu}$.

Using the 4-momenta $P_{A,\mu}$, q_μ, $P^{sp}_{A-1,\mu}$, we can construct the tensor $\hat{W}^{IA}_{\mu\nu}$ satisfying the condition of gauge invariance and symmetry as follows

$$\hat{W}^{IA}_{\mu\nu} = \left(-g_{\mu\nu} + \frac{q_\mu q_\nu}{q^2}\right)\hat{W}^{IA}_1 + \left(P_{A,\mu} - \frac{\nu_A}{q^2} q_\mu\right)\left(P_{A,\nu} - \frac{\nu_A}{q^2} q_\nu\right)\hat{W}^{IA}_2 + \\ + \left(P^{sp}_{A-1,\mu} - \frac{\nu_{sp}}{q^2} q_\mu\right)\left(P^{sp}_{A-1,\nu} - \frac{\nu_{sp}}{q^2} q_\nu\right)\hat{W}^{IA}_3 + \\ + \frac{1}{2}\left[\left(P_{A,\mu} - \frac{\nu_A}{q^2} q_\mu\right)\left(P^{sp}_{A-1,\nu} - \frac{\nu_{sp}}{q^2} q_\nu\right) + \left(P_{A,\nu} - \frac{\nu_A}{q^2} q_\nu\right)\left(P^{sp}_{A-1,\mu} - \frac{\nu_{sp}}{q^2} q_\mu\right)\right]\hat{W}^{IA}_4 \tag{5.8}$$

From the self-adjointness of the hadronic electromagnetic current J_μ it follows that the structure functions $\hat{W}_i^{lA}$ are real scalar functions of the kinematic invariants q^2, ν_A, ν_{sp}, x:

$$\hat{W}_i^{lA} = \hat{W}_i^{lA}(q^2, \nu_A, \nu_{sp}, x) \; ; \quad i = 1, \ldots, 4 .$$

When the cross section (5.4) is integrated over the azimuthal angle φ_{sp} of the spectator fragment, the integration affects only the tensor $\hat{W}_{\mu\nu}^{lA}$. This integration over the angle φ_{sp} makes it possible to define new structure functions $\tilde{W}_i^{lA}$ in the following manner

$$\tilde{W}_{\mu\nu}^{lA} = \int_0^{2\pi} d\varphi_{sp} \hat{W}_{\mu\nu}^{lA} =$$

$$= \int d\varphi_{sp} d\nu_{sp} dx \, \delta(\nu_{sp} - P_{A-1}^{sp} q) \, \delta(x - P_A P_{A-1}^{sp}) \hat{W}_{\mu\nu}^{lA} = \tag{5.9}$$

$$= \left(-g_{\mu\nu} + \frac{q_\mu q_\nu}{q^2}\right) \tilde{W}_1^{lA} + \left(P_{A,\mu} - \frac{\nu_A}{q^2} q_\mu\right)\left(P_{A,\nu} - \frac{\nu_A}{q^2} q_\nu\right) \tilde{W}_2^{lA} .$$

The structure functions $\tilde{W}_i^{lA}$ as well as the functions $\hat{W}_i^{lA}$ depend on the invariants q^2, ν_A, ν_{sp}, x:

$$\tilde{W}_i^{lA} = \tilde{W}_i^{lA}(q^2, \nu_A, \nu_{sp}, x) .$$

Using the new structure functions $\tilde{W}_i^{lA}$, the cross section (5.4) can be rewritten in the form

$$\frac{d\sigma(lA \to l'(A-1)X)}{dq^2 d\nu_A d\nu_{sp} dx} = \frac{(\nu_A^2 - m_A^2 q^2)^{-1/2}}{8(2\pi)^5 \lambda(s_{lA}, m_l^2, m_A^2)} \left(\frac{4\pi\alpha}{q^2}\right)^2$$

$$\left\{ -2(q^2 + 2m_l^2) \tilde{W}_1^{lA}(q^2, \nu_A, \nu_{sp}, x) + \right. \tag{5.10}$$

$$\left. + \left[(s_{lA} - \nu_A - m_A^2 - m_l^2)^2 + q^2\left(m_A^2 - \frac{\nu_A^2}{q^2}\right)\right] \tilde{W}_2^{lA}(q^2, \nu_A, \nu_{sp}, x) \right\}$$

Consider now the process of absorption of the virtual photon with four-momentum q on the nucleus A when the spectator fragment $(A-1)$ is detected in the final state. The cross section of this process is of the form

$$\frac{d\sigma(\gamma^* A \to (A-1)X)}{d\vec{P}^{sp}_{A-1}/E^{sp}_{A-1}} = \frac{4\pi^2\alpha}{\nu_A + q^2/2}\varepsilon^*_\mu \varepsilon_\nu \hat{W}^{1A}_{\mu\nu}. \tag{5.11}$$

$\hat{W}^{1A}_{\mu\nu}$ is the tensor defined above (see Eq. (5.6)), ε_μ is the virtual photon polarization four-vector, satisfying the gauge condition $\varepsilon_\mu q_\mu = 0$. In the reference frame, in which $q = (q_0, 0, 0, q_z)$, the polarization vector ε_μ can be chosen in the form

$$\varepsilon^{T,\pm}_\mu = \frac{1}{\sqrt{2}}(0, 1, \pm i, 0) \text{ - transversal polarization}$$

$$\varepsilon^{L}_\mu = \frac{1}{\sqrt{-q^2}}(q_z, 0, 0, q_0) \text{ - longitudinal polarization} \tag{5.12}$$

Integrating the cross section (5.11) over the azimuthal angle φ_{sp} of the spectator fragment and taking into account the Eq. (5.9), we obtain the following expression for cross section of the deep inelastic inclusive photoabsorption on the nucleus:

$$\left\langle \frac{d\sigma(\gamma^* A \to (A-1)X)}{d\vec{P}^{sp}_{A-1}/E^{sp}_{A-1}} \right\rangle = \frac{4\pi^2\alpha}{\nu_A + q^2/2}\varepsilon^*_\mu \varepsilon_\nu \tilde{W}^{1A}_{\mu\nu} \tag{5.13}$$

Here and in what follows the symbol $\langle\ \rangle$ denotes that integration over the azimuthal angle φ_{sp} of the spectator fragment is carried out in the corresponding cross section.

Taking into account the form of the tensor $\tilde{W}^{1A}_{\mu\nu}$ we obtain the following expression for the photoabsorption cross section of transversally and longitudinally polarized photons:

$$\left\langle \frac{d\sigma^{T}(\gamma^{*}A\to(A-1)X)}{d\vec{P}^{sp}_{A-1}/E^{sp}_{A-1}} \right\rangle = \frac{4\pi^2\alpha}{\nu_A+q^2/2}\,\tilde{W}^{lA}_1(q^2,\nu_A,\nu_{sp},x) \tag{5.14a}$$

$$\left\langle \frac{d\sigma^{L}(\gamma^{*}A\to(A-1)X)}{d\vec{P}^{sp}_{A-1}/E^{sp}_{A-1}} \right\rangle = \frac{4\pi^2\alpha}{\nu_A+q^2/2}\Big\{-\tilde{W}^{lA}_1(q^2,\nu_A,\nu_{sp},x)+$$

$$+\left(m^2_A-\frac{\nu^2_A}{q^2}\right)\tilde{W}^{lA}_2(q^2,\nu_A,\nu_{sp},x)\Big\} \tag{5.14b}$$

Derive now a relation between the cross sections of the processes $lA\to l'(A-1)X$ and $\gamma A\to(A-1)X$ in the limit $q^2\to 0$. The cross section of photoabsorption of the longitudinally polarized photons should vanish in this limit as it corresponds to the real photons which have transversal polarization only. This leads to the relation

$$\lim_{q^2\to 0}\tilde{W}^{lA}_1(q^2,\nu_A,\nu_{sp},x)=$$

$$=\lim_{q^2\to 0}\left[\left(m^2_A-\frac{\nu^2_A}{q^2}\right)\tilde{W}^{lA}_2(q^2,\nu_A,\nu_{sp},x)\right]= \tag{5.15}$$

$$=\lim_{q^2\to 0}\left[(\nu^2_A/q^2)\,\tilde{W}^{lA}_2(q^2,\nu_A,\nu_{sp},x)\right]$$

In the limit $q^2\to 0$ the cross section of the transversally polarized photons is the total photoabsorption cross section and takes the form

$$\left.\frac{d\sigma(\gamma A\to(A-1)X)}{d\nu_{sp}\,dx}\right|_{q^2\to 0}=\frac{4\pi^2\alpha}{\nu^2_A}\lim_{q^2\to 0}\tilde{W}^{lA}_1(q^2,\nu_A,\nu_{sp},x) \tag{5.16}$$

Taking into account (5.15), we have

$$\left.\frac{d\sigma(\gamma A\to(A-1)X)}{d\nu_{sp}\,dx}\right|_{q^2\to 0}\simeq -4\pi^2\alpha\lim_{q^2\to 0}\left[\frac{1}{q^2}\tilde{W}^{lA}_2(q^2,\nu_A,\nu_{sp},x)\right] \tag{5.17}$$

To establish the relation of this cross section to the cross section of deep inelastic electroproduction let us write (5.10) in the laboratory frame (the rest frame of the nucleus A). Neglecting the lepton mass, we get:

$$\frac{d\sigma(lA \to l'(A-1)X}{dq^2 d\nu_A d\nu_{sp} dx} = \frac{1}{8(2\pi)^5 m_A^2(\nu_A^2 - m_A^2 q^2)^{1/2}}\left(\frac{E_{k'}}{E_k}\right)\left(\frac{4\pi\alpha}{q^2}\right)^2 \times$$
$$\times \left\{2\sin^2\frac{\vartheta}{2}\, \tilde{W}_1^{lA}(q^2, \nu_A, \nu_{sp}, x) + m_A^2 \cos^2\frac{\vartheta}{2}\, \tilde{W}_2^{lA}(q^2, \nu_A, \nu_{sp}, x)\right\} \tag{5.18}$$

Here E_k and $E_{k'}$ are the energies of the initial and final leptons, respectively, ϑ is the scattering angle (the angle between the directions of momenta $\vec{k}$ and $\vec{k}'$ of the initial and final leptons).

For $q^2 \to 0$ we have $\sin^2\frac{\vartheta}{2} \to 0$, and hence, $\cos^2\frac{\vartheta}{2} \to 1$ and from the cross section (5.18) we can obtain

$$\left.\frac{d\sigma(lA \to l'(A-1)X)}{dq^2 d\nu_A d\nu_{sp} dx}\right|_{q^2 \to 0} =$$
$$= \frac{\alpha^2}{2(2\pi)^3 \nu_A}\left(\frac{E_{k'}}{E_k}\right) \lim_{q^2 \to 0}\left[\frac{1}{q^4}\tilde{W}_2^{lA}(q^2, \nu_A, \nu_{sp}, x)\right] \tag{5.19}$$

The Equations (5.17) and (5.19) yield the following relation between the cross sections of electroproduction and photoabsorption on the nucleus:

$$\left.\frac{d\sigma(lA \to l'(A-1)X)}{dq^2 d\nu_A d\nu_{sp} dx}\right|_{q^2 \to 0} \simeq \tag{5.20}$$
$$\simeq \frac{\alpha}{2(2\pi)^5 \nu_A}\left(\frac{E_{k'}}{E_k}\right)\left(\frac{1}{q^2}\right)\frac{d\sigma(\gamma A \to (A-1)X)}{d\nu_{sp} dx}$$

2. Deep Inelastic Scattering of the Neutrino (Antineutrino) on the Nucleus. The Process $\nu A \to l(A-1)X$

Consider now the charged current induced deep inelastic neutrino-nucleus interactions in the framework of the V-A-theory of weak interactions. Define the kinematic invariants of this process similar to the case of deep inelastic electroproduction (see Eq. (5.1) and (5.2)), but now let k and k' be the four-momenta of the initial neutrino (antineutrino) and the final lepton, respectively.

In the lowest order in the weak interaction the matrix element of the deep inelastic interaction of the neutrino (antineutrino) with the nucleus is of the form

$$T(\nu A \to l(A-1)X) = \frac{G}{\sqrt{2}} l_\mu^\pm \langle P_{A-1}^{sp}, X | J_\mu^\pm(0) | P_A \rangle \qquad (5.21)$$

here G is the weak interaction constant ($Gm_p^2 = 1{,}02 \cdot 10^5$), $J_\mu^\pm$ are weak hadronic charged currents, $l_\mu^\pm$ are the corresponding weak leptonic currents:

$$\begin{aligned} \nu_l \to l: \quad & l_\mu^- = \bar{u}(k')\gamma_\mu(1+\gamma_5)u(k) \\ \bar{\nu}_l \to \bar{l}: \quad & l_\mu^+ = \bar{v}(k')\gamma_\mu(1-\gamma_5)v(k) \end{aligned} \qquad (5.22)$$

When the lepton l and the spectator nucleus $(A-1)$ are detected in the final state, the cross section of the deep inelastic neutrino (antineutrino)-nucleus interaction looks as follows:

$$\frac{d\sigma(\nu A \to l(A-1)X)}{dq^2 d\nu_A d\nu_{sp} dx d\varphi_{sp}} = \frac{G^2}{16(2\pi)^5(s_{\nu A} - m_A^2)^2} \times$$
$$\times (\nu_A^2 - m_A^2 q^2)^{-1/2} m_{\mu\nu}^{\nu,\bar{\nu}} \hat{W}_{\mu\nu}^{\nu,\bar{\nu}}. \qquad (5.23)$$

The tensor $m_{\mu\nu}^{\nu,\bar{\nu}}$ defines the leptonic part of the process and is the product of the weak leptonic currents averaged over the spin states of the neutrino (antineutrino) and summed over the spin states of the final lepton

$$m_{\mu\nu}^{\nu,\bar{\nu}} = \overline{\sum_{spin}} (l_\mu^{\mp})^* l_\nu^{\mp} = 8[k_\mu k'_\nu + k_\nu k'_\mu - g_{\mu\nu}(kk') \pm i\varepsilon_{\mu\nu\lambda\rho} k_\lambda k'_\rho] \qquad (5.24)$$

The strong interaction part of the process is described by the tensor $\hat{W}_{\mu\nu}^{\nu,\bar{\nu}}$ which is defined as

$$\hat{W}_{\mu\nu}^{\nu,\bar{\nu}}(P_A, q, P_{A-1}^{sp}) = \overline{\sum_{spin}} \sum_X (2\pi)^4 \delta^{(4)}(P_A + q - P_{A-1}^{sp} - P_X) \times \\ \times \langle P_A | (J_\mu^{\pm}(0))^* | P_{A-1}^{sp}, X \rangle \langle P_{A-1}^{sp}, X | J_\nu^{\pm}(0) | P_A \rangle \qquad (5.25)$$

In contrast to the case of deep inelastic electroproduction on nuclei, the requirements of symmetry and gauge invariance are not imposed on the tensor $\hat{W}_{\mu\nu}^{\nu,\bar{\nu}}$. Hence, the most general form of this tensor which is constructed by means of the momenta $P_{A,\mu}$, q_μ and $P_{A-1,\mu}^{sp}$ is as follows:

$$\hat{W}_{\mu\nu}^{\nu,\nu} = \left(-g_{\mu\nu} + \frac{q_\mu q_\nu}{q^2}\right)\hat{W}_1^{\nu,\nu} + \left(P_{A,\mu} - \frac{\nu_A}{q^2} q_\mu\right)\left(P_{A,\nu} - \frac{\nu_A}{q^2} q_\nu\right)\hat{W}_2^{\nu,\bar{\nu}} - \\ - \frac{i}{2}\varepsilon_{\mu\nu\lambda\rho} P_{A,\lambda} q_\rho \hat{W}_3^{\nu,\bar{\nu}} + q_\mu q_\nu \hat{W}_4^{\nu,\bar{\nu}} + (P_{A,\mu} q_\nu + P_{A,\nu} q_\mu)\hat{W}_5^{\nu,\bar{\nu}} + \\ + i(P_{A,\mu} q_\nu + P_{A,\nu} q_\mu)\hat{W}_6^{\nu,\bar{\nu}} + \left(P_{A-1,\mu}^{sp} - \frac{\nu_{sp}}{q^2} q_\mu\right)\left(P_{A-1,\nu}^{sp} - \frac{\nu_{sp}}{q^2} q_\nu\right)\hat{W}_7^{\nu,\bar{\nu}} + \\ + \frac{1}{2}\left[\left(P_{A,\mu} - \frac{\nu_A}{q^2} q_\mu\right)\left(P_{A-1,\nu}^{sp} - \frac{\nu_{sp}}{q^2} q_\nu\right) + \left(P_{A,\nu} - \frac{\nu_A}{q^2} q_\nu\right)\left(P_{A-1,\mu}^{sp} - \frac{\nu_{sp}}{q^2} q_\mu\right)\right]\hat{W}_8^{\nu,\bar{\nu}} - \\ - \frac{i}{2}\varepsilon_{\mu\nu\lambda\rho} P_{A-1,\lambda}^{sp} q_\rho \hat{W}_9^{\nu,\bar{\nu}} + (P_{A-1,\mu}^{sp} q_\nu + P_{A-1,\nu}^{sp} q_\mu)\hat{W}_{10}^{\nu,\bar{\nu}} + \qquad (5.26) \\ + i(P_{A-1,\mu}^{sp} q_\nu - P_{A-1,\nu}^{sp} q_\mu)\hat{W}_{11}^{\bar{\nu},\nu} - \frac{i}{2}\varepsilon_{\mu\nu\lambda\rho} P_{A,\lambda} P_{A-1,\rho}^{sp} \hat{W}_{12}^{\bar{\nu},\nu} + \\ + i(P_{A,\mu} P_{A-1,\nu}^{sp} - P_{A,\nu} P_{A-1,\mu}^{sp})\hat{W}_{13}^{\nu,\bar{\nu}}$$

From the conditions of the hermitean conjugation of the weak hadronic currents $J_\mu^{\pm}$ it follows that the structure functions $\hat{W}_i^{\nu,\bar{\nu}}$ are real scalar functions of the kinematic invariants q^2, ν_A, ν_{sp} and x :

$$\hat{W}_i^{\nu,\bar{\nu}} = \hat{W}_i^{\nu,\bar{\nu}}(q^2, \nu_A, \nu_{sp}, x)$$

The time-reflection symmetry implies

$$\hat{W}_{\mu\nu}^{\nu,\bar{\nu}}(P_A, q, P_{A-1}^{sp}) = \pm\left[\hat{W}_{\mu\nu}^{\nu,\bar{\nu}}(\tilde{P}_A, \tilde{q}, \tilde{P}_{A-1}^{sp})\right]^{+} \tag{5.27}$$

"+" corresponds to $\mu=\nu=0$, or $\mu,\nu\neq 0$

"-" corresponds to $\mu=0$, $\nu\neq 0$ or $\mu\neq 0$, $\nu=0$ and

$$\tilde{P}_A = (E_A, -\vec{P}_A), \qquad \tilde{q} = (q_0, -\vec{q}), \qquad \tilde{P}_{A-1}^{sp} = (E_{A-1}^{sp}, -\vec{P}_{A-1}^{sp})$$

From the Eqs. (5.26) and (5.27) follows

$$\hat{W}_6^{\nu,\bar{\nu}} = \hat{W}_{11}^{\nu,\bar{\nu}} = \hat{W}_{13}^{\nu,\bar{\nu}} = 0 .$$

When the cross section (5.23) is integrated over the azimuthal angle φ_{sp} of the spectator fragment, the integration affects only the tensor $\hat{W}_{\mu\nu}^{\nu,\bar{\nu}}$. This integration makes it possible to introduce new structure functions $\tilde{W}_i^{\nu,\bar{\nu}}$ as follows:

$$\begin{aligned}\tilde{W}_{\mu\nu}^{\nu,\bar{\nu}} &= \int_0^{2\pi} d\varphi_{sp}\, \hat{W}_{\mu\nu}^{\nu,\bar{\nu}} = \\ &= \int d\varphi_{sp}\, d\nu_{sp}\, dx\, \delta(\nu_{sp} - P_{A-1}^{sp} q)\, \delta(x - P_A P_{A-1}^{sp})\, \hat{W}_{\mu\nu}^{\nu,\bar{\nu}} = \\ &= \left(-g_{\mu\nu} + \frac{q_\mu q_\nu}{q^2}\right)\tilde{W}_1^{\nu,\bar{\nu}} + \left(P_{A,\mu} - \frac{\nu_A}{q^2} q_\mu\right)\left(P_{A,\nu} - \frac{\nu_A}{q^2} q_\nu\right)\tilde{W}_2^{\nu,\bar{\nu}} - \\ &- \frac{i}{2}\varepsilon_{\mu\nu\lambda\rho} P_{A,\lambda} q_\rho \tilde{W}_3^{\nu,\bar{\nu}} + q_\mu q_\nu \tilde{W}_4^{\nu,\bar{\nu}} + (P_{A,\mu} q_\nu + P_{A,\nu} q_\mu)\tilde{W}_5^{\nu,\bar{\nu}} .\end{aligned} \tag{5.28}$$

Using the new structure functions we obtain the following expression for the cross section of the deep inelastic neutrino (antineutrino)-nucleus scattering :

$$\frac{d\sigma(\nu A\to l(A-1)X)}{dq^2 d\nu_A d\nu_{sp} d\varkappa} = \frac{G^2}{4(2\pi)^3(s_{\nu A}-m_A^2)^2}(\nu_A^2-m_A^2q^2)^{-1/2}$$

$$\Big\{-\frac{1}{q^2}(2q^2+m_l^2)(q^2-m_l^2)\widetilde{W}_1^{\nu,\bar\nu}(q^2,\nu_A,\nu_{sp},\varkappa)+$$

$$+\Big[\Big(s_{lA}-\nu_A-m_A^2+m_l^2\frac{\nu_A}{q^2}\Big)^2+(q^2-m_l^2)\Big(m_A^2-\frac{\nu_A^2}{q^2}\Big)\Big]\widetilde{W}_2^{\nu,\bar\nu}(q^2,\nu_A,\nu_{sp},\varkappa)\pm \quad (5.29)$$

$$\pm[(s_{lA}-\nu_A-m_A^2)q^2+\nu_A m_A^2]\widetilde{W}_3^{\nu,\bar\nu}(q^2,\nu_A,\nu_{sp},\varkappa)-$$

$$-m_l^2(q^2-m_l^2)\widetilde{W}_4^{\nu,\bar\nu}(q^2,\nu_A,\nu_{sp},\varkappa)-$$

$$-2m_l^2(s_{lA}-m_A^2)\widetilde{W}_5^{\nu,\bar\nu}(q^2,\nu_A,\nu_{sp},\varkappa)\Big\}$$

The sign "+" before the third term in this expression corresponds to the scattering of the neutrino, while the sign "-" corresponds to the scattering of the antineutrino. It can be seen from (5.29) that the contributions of the structure functions $\widetilde{W}_4^{\nu,\bar\nu}$ and $\widetilde{W}_5^{\nu,\bar\nu}$ vanish if the lepton mass is neglected.

If we assume that the interaction between the neutrino (or antineutrino) and the nucleus takes place through the exchange of a virtual $W^{\pm}$-boson, then in this case the cross section of the deep inelastic neutrino (antineutrino)-nucleus scattering will differ from (5.29) by the factor $[1/(1-q^2/m_W^2)]^2$, which tends to unity when $m_W^2 \gg q^2$ (m_W is the mass of the intermediate W-boson).

Consider now the absorption of the massless particle with quantum numbers of the $W^{\pm}$-boson by the nucleus A when the spectator fragment (A-1) of the initial nucleus A is detected and the system of hadrons X is produced in the final state. The cross section of this process can be written in the form

$$\frac{d\sigma(WA\to(A-1)X)}{d\vec{P}_{A-1}^{sp}/E_{A-1}^{sp}} = \frac{g^2}{(2\pi)^3(\nu_A+q^2/2)}\varepsilon_\mu^*\varepsilon_\nu\hat{W}_{\mu\nu}^{\nu,\bar\nu} \quad (5.30)$$

The tensor $\hat{W}_{\mu\nu}^{\nu,\bar\nu}$ is defined by Eq. (5.26), ε_μ is the normalized polarization four-vector which satisfies the condition $\varepsilon_\mu q_\mu = 0$. In the reference frame, in which $q=(q_0,0,0,q_z)$, the polarization vector ε_μ can be chosen in the form:

$$\varepsilon_\mu^R = \frac{1}{\sqrt{2}}(0,1,i,0) \quad \text{- right-hand polarization,}$$

$$\varepsilon_\mu^L = \frac{1}{\sqrt{2}}(0,i,1,0) \quad \text{- left-hand polarization,} \tag{5.31}$$

$$\varepsilon_\mu^S = \frac{1}{\sqrt{-q^2}}(q_z,0,0,q_0) \quad \text{- scalar polarization.}$$

Integrating the cross section (5.30) over the azimuthal angle φ_{sp} of the spectator fragment and taking into account definition (5.28), we obtain the following expressions for the cross sections of the massless W-boson absorption

$$\left\langle \frac{d\sigma^R(WA\to(A-1)X)}{d\vec{P}^{sp}_{A-1}/E^{sp}_{A-1}} \right\rangle = \frac{\sigma_0}{2(2\pi)^4}\Big\{\tilde{W}_1^{\nu,\bar{\nu}}(q^2,\nu_A,\nu_{sp},x) + \frac{1}{2}\Big(1-\frac{m_A^2q^2}{\nu_A^2}\Big)^{1/2}\nu_A\tilde{W}_3^{\nu,\bar{\nu}}(q^2,\nu_A,\nu_{sp},x)\Big\}, \tag{5.32a}$$

$$\left\langle \frac{d\sigma^L(WA\to(A-1)X)}{d\vec{P}^{sp}_{A-1}/E^{sp}_{A-1}} \right\rangle = \frac{\sigma_0}{2(2\pi)^4}\Big\{\tilde{W}_1^{\nu,\bar{\nu}}(q^2,\nu_A,\nu_{sp},x) - \frac{1}{2}\Big(1-\frac{m_A^2q^2}{\nu_A^2}\Big)^{1/2}\nu_A\tilde{W}_3^{\nu,\bar{\nu}}(q^2,\nu_A,\nu_{sp},x)\Big\}, \tag{5.32b}$$

$$\left\langle \frac{d\sigma^S(WA\to(A-1)X)}{d\vec{P}^{sp}_{A-1}/E^{sp}_{A-1}} \right\rangle = \frac{\sigma_0}{2(2\pi)^4}\Big\{-\tilde{W}_1^{\nu,\bar{\nu}}(q^2,\nu_A,\nu_{sp},x) + \Big(1-\frac{\nu_A^2}{m_A^2q^2}\Big)m_A^2\tilde{W}_2^{\nu,\bar{\nu}}(q^2,\nu_A,\nu_{sp},x)\Big\}. \tag{5.32c}$$

Here $\sigma_0 = \pi g^2/(\nu_A + q^2/2)$.

3. Lepton Pair Production in Hadron-Nucleus Collisions. The Process $aA \to l^+l^-(A-1)X$.

Consider the hadron-nucleus interaction when the lepton pair (e.g., the muon pair $\mu^+\mu^-$) and the spectator fragment $(A-1)$ are detected in the final state. Let p_a and P_A be four-momenta of the colliding hadron and nucleus, respectively, k and k' be the four-momenta of the created leptons, and P^{sp}_{A-1} be the four-momentum of the spectator fragment. The kinematical invariants of the process $aA \to l^+l^-(A-1)X$ are defined as follows:

$$
\begin{gathered}
q^2=(k+k')^2, \quad \nu_A=P_A q, \quad \nu_a=p_a q, \quad \nu_{sp}=P^{sp}_{A-1}q, \\
\varkappa_A=P_A P^{sp}_{A-1}, \quad \varkappa_a=p_a P^{sp}_{A-1}, \quad s_{aA}=(p_a+P_A)^2.
\end{gathered}
\tag{5.33}
$$

In the lowest order in the electromagnetic interaction the matrix element of the process is of the form

$$
T(aA \to l^+l^-(A-1)X)=\left(\frac{4\pi\alpha}{q^2}\right)\varepsilon_\mu \langle P^{sp}_{A-1}, X; \mathrm{out}| J_\mu(0)|P_A, p_a; \mathrm{in}\rangle \tag{5.34}
$$

Here ε_μ is the electromagnetic current of the lepton pair $\varepsilon_\mu=\bar{u}(k)\gamma_\mu v(k')$.

The cross section of the process can be written as

$$
\begin{gathered}
d\sigma(aA \to l^+l^-(A-1)X)=\frac{1}{16(2\pi)^9\lambda^{1/2}(s_{aA}, m_a^2, m_A^2)}\left(\frac{4\pi\alpha}{q^2}\right)^2 \times \\
\times\, \varepsilon^*_\mu \varepsilon_\nu \rho_{\mu\nu}\left(d\vec{k}/E_k\right)\left(d\vec{k}'/E_{k'}\right)\left(d\vec{P}^{sp}_{A-1}/E^{sp}_{A-1}\right)
\end{gathered}
\tag{5.35}
$$

m_a and m_A are the masses of the colliding hadron and nucleus, respectively.

Summing this cross section over the spin states of the lepton pair, it can be rewritten in the form

$$d\sigma(aA\to l^+l^-(A-1)X)=\frac{1}{4(2\pi)^3\lambda^{1/2}(s_{aA},m_a^2,m_A^2)}\left(\frac{4\pi\alpha}{q^2}\right)^2$$

$$\pi(q^2)(-g_{\mu\nu}q^2+q_\mu q_\nu)\rho_{\mu\nu}\left(\frac{d^4q}{(2\pi)^4}\right)\left(\frac{d\vec{P}^{sp}_{A-1}}{E^{sp}_{A-1}}\right) \tag{5.36}$$

where the leptonic part of the process is defined by the expression

$$\pi(q^2)(-g_{\mu\nu}q^2+q_\mu q_\nu)=$$

$$=\sum_{spin}\int\frac{d\vec{k}d\vec{k}'}{(2\pi)^6 2E_k 2E_{k'}}(2\pi)^4\delta^{(4)}(q-k-k')\varepsilon^*_\mu\varepsilon_\nu=$$

$$=\left(1-\frac{4m_l^2}{q^2}\right)^{1/2}\int\frac{d\Omega}{8\pi^2}\left[k_\mu k'_\nu+k_\mu k'_\nu-\frac{1}{2}g_{\mu\nu}(k+k')^2\right] \tag{5.37}$$

m_l is the lepton (muon) mass, $d\Omega$ is the element of the solid angle which is related to the direction of one of the produced muons in their center-of-mass frame ($\vec{q}=\vec{k}+\vec{k}'=0$). Here

$$\vec{k}=-\vec{k}'=|\vec{k}|(\sin\vartheta\cos\varphi,\sin\vartheta\sin\varphi,\cos\vartheta)$$

$$E_k=E_{k'}=\sqrt{\vec{k}^2+m_l^2}=\frac{1}{2}\sqrt{q^2}$$

After the angular integration we have

$$\pi(q^2)=\frac{1}{6\pi}\left(1+\frac{2m_l^2}{q^2}\right)\left(1-\frac{4m_l^2}{q^2}\right)^{1/2}\Bigg|_{m_l=0}=\frac{1}{6\pi} \tag{5.38}$$

In Eqs. (5.35) and (5.36) the tensor $\rho_{\mu\nu}$ describes the strongly interacting part of the process and is defined by

$$\rho_{\mu\nu}(P_A,p_a,q,P^{sp}_{A-1})=\overline{\sum_{spin}}\sum_X(2\pi)^4\delta^{(4)}(P_A+p_a-q-P^{sp}_{A-1}-p_X)\times$$

$$\times\langle P_A,p_a;in|J_\mu(0)|P^{sp}_{A-1},X;out\rangle\langle P^{sp}_{A-1},X;out|J_\nu(0)|P_A,p_a;in\rangle \tag{5.39}$$

Here the summation over the all final states with the spectator fragment is performed and for each of these states the summation over the spin states and integration over the phase spaces of the undetected final hadrons are carried out.

The conservation of the electromagnetic current J_μ leads to the following condition of gauge invariance for $\rho_{\mu\nu}$:

$$q_\mu \rho_{\mu\nu} = \rho_{\mu\nu} q_\nu = 0 \tag{5.40}$$

From the self-adjointness of J_μ follows that the real part of this tensor must be symmetric and the imaginary part must be antisymmetric under the $\mu \to \nu$ exchange. The most general form of the tensor $\rho_{\mu\nu}$ constructed by means of the four-vectors $P_{A,\mu}$, $p_{a,\mu}$, q_μ and $P^{sp}_{A-1,\mu}$ is

$$\begin{aligned}\rho_{\mu\nu} = &(-g_{\mu\nu} + q_\mu q_\nu/q^2)\rho_1 + \mathcal{P}_{A,\mu}\mathcal{P}_{A,\nu}\rho_2 + \mathcal{P}_{a,\mu}\mathcal{P}_{a,\nu}\rho_3 + \\ &+ (\mathcal{P}_{A,\mu}\mathcal{P}_{a,\nu} + \mathcal{P}_{A,\nu}\mathcal{P}_{a,\mu})\rho_4 + i(\mathcal{P}_{A,\mu}\mathcal{P}_{a,\nu} - \mathcal{P}_{A,\nu}\mathcal{P}_{a,\mu})\rho_5 + \\ &+ \mathcal{P}^{sp}_{A-1,\mu}\mathcal{P}^{sp}_{A-1,\nu}\rho_6 + (\mathcal{P}_{A,\mu}\mathcal{P}^{sp}_{A-1,\nu} + \mathcal{P}_{A,\nu}\mathcal{P}^{sp}_{A-1,\mu})\rho_7 + \\ &+ i(\mathcal{P}_{A,\mu}\mathcal{P}^{sp}_{A-1,\nu} - \mathcal{P}_{A,\nu}\mathcal{P}^{sp}_{A-1,\mu})\rho_8 + (\mathcal{P}_{a,\mu}\mathcal{P}^{sp}_{A-1,\nu} + \mathcal{P}_{a,\nu}\mathcal{P}^{sp}_{A-1,\mu})\rho_9 + \\ &+ i(\mathcal{P}_{a,\mu}\mathcal{P}^{sp}_{A-1,\nu} - \mathcal{P}_{a,\nu}\mathcal{P}^{sp}_{A-1,\mu})\rho_{10}\end{aligned} \tag{5.41}$$

Here $\mathcal{P}_{A,\mu} = P_{A,\mu} - \frac{\nu_A}{q^2}q_\mu$, $\quad \mathcal{P}_{a,\mu} = p_{a,\mu} - \frac{\nu_a}{q^2}q_\mu$, $\quad \mathcal{P}^{sp}_{A-1,\mu} = P^{sp}_{A-1,\mu} - \frac{\nu_{sp}}{q^2}q_\mu$.

Due to the symmetry of the leptonic part (see eq. (5.37)) of the process considered a contribution to the cross section is given only by the symmetric part of the tensor $\rho_{\mu\nu}$. The structure functions ρ_i are real scalar functions of the kinematic invariants: $\rho_i = \rho_i(s_{aA}, q^2, \nu_A, \nu_a, \nu_{sp}, x_A, x_a)$.
Expressing the phase spaces of the detected particles in terms of the invariants

$$d^4q = \frac{dq^2 d\nu_A d\nu_a d\varphi_q}{\lambda^{1/2}(s_{aA}, m_a^2, m_A^2)}; \quad \frac{d\vec{P}^{sp}_{A-1}}{E^{sp}_{A-1}} = 2\frac{dx_A dx_a d\varphi_{sp}}{\lambda^{1/2}(s_{aA}, m_a^2, m_A^2)} \tag{5.42}$$

we obtain the following expression for the differential cross section of the process $aA \to l^+l^-(A-1)X$

$$\frac{d\sigma(aA\to l^+l^-(A-1)X)}{dq^2 d\nu_A d\nu_a d\varphi_q dx_A dx_a d\varphi_{sp}} = \frac{1}{2(2\pi)^7 \lambda^{3/2}(s_{aA}, m_a^2, m_A^2)} \left(\frac{4\pi\alpha}{q^2}\right)^2 \times$$
$$\times \pi(q^2)(-g_{\mu\nu}q^2 + q_\mu q_\nu)\rho_{\mu\nu} \tag{5.43}$$

The arguments of the structure functions ρ_i depend on the azimuthal angles of the virtual photon and the spectator fragment. It is impossible therefore to integrate the cross section (5.43) over the azimuthal angles in the general form because the structure functions are unknown. To perform the integration consider a particular case when the colliding particles (the hadron a and the nucleus A) move along the z-axis and the total transverse momentum of the lepton pair is zero ($\vec{q}_\perp=0$). In this case the structure functions ρ_i do not depend on the azimuthal angles φ_q and φ_{sp} and we can introduce new structure functions $\tilde{\rho}_i$ by the integration of tensor $\rho_{\mu\nu}$ over the angle φ_{sp}:

$$\tilde{\rho}_{\mu\nu} = \int d\varphi_{sp}\, \rho_{\mu\nu} =$$
$$= \left(-g_{\mu\nu} - \frac{q_\mu q_\nu}{q^2}\right)\tilde{\rho}_1 + P_{A,\mu}P_{A,\nu}\tilde{\rho}_2 + P_{a,\mu}P_{a,\nu}\tilde{\rho}_3 +$$
$$+ (P_{A,\mu}P_{a,\nu} + P_{A,\nu}P_{a,\nu})\tilde{\rho}_4 + i(P_{A,\mu}P_{a,\nu} - P_{A,\nu}P_{a,\mu})\tilde{\rho}_5$$

Then for the cross section of the process $aA \to l^+l^-(A-1)X$ we obtain

$$\frac{d\sigma(aA\to l^+l^-(A-1)X)}{dq^2 d\nu_A d\nu_a dx_A dx_a} = \frac{1}{2(2\pi)^6 \lambda^{3/2}(s_{aA}, m_a^2, m_A^2)} \times$$
$$\times \left(\frac{4\pi\alpha}{q^2}\right)^2 q^2 \pi(q^2) \left\{3\tilde{\rho}_1 - \left(m_A^2 - \frac{\nu_A^2}{q^2}\right)\tilde{\rho}_2 - \right. \tag{5.44}$$
$$\left. - \left(m_a^2 - \frac{\nu_a^2}{q^2}\right)\tilde{\rho}_3 - \left(s_{aA} - m_A^2 - m_a^2 - 2\frac{\nu_A \nu_a}{q^2}\right)\tilde{\rho}_4 \right\}$$

Note that in particular case $(P_A=(E_A,0,0,P_{A,z}),\ p_a=(E_a,0,0,p_{a,z}),\ q=(q_0,0,0,q_z))$ not all the arguments of the structure functions $\tilde{\rho}_i$ are independent. The invariant ν_{sp}, e.g., can be expressed by means of other invariants in the following manner:

$$\nu_{sp}=2\lambda^{-1}(s_{aA},m_A^2,m_a^2)\{s_{aA}(x_A\nu_a+x_a\nu_A)+$$
$$+(x_A-x_a)(\nu_A m_a^2-\nu_a m_A^2)+(\nu_A-\nu_a)(x_A m_a^2-x_a m_A^2)\}$$

Thus the number of independent kinematic variables on which the structure functions $\tilde{\rho}_i$ depend is reduced.

4. Dimensional Analysis and the Principle of Automodelity

Let us carry out now the dimensional analysis of the structure functions introduced in the preceding sections. For the case of the deep inelastic lepton-nucleus scattering we shall work in the centre-of-mass frame of the nucleus A and of the virtual intermediate particle in the limit $P_{A,z}\to\infty$ $(P_A=(E_A,0,0,P_{A,z}),\ q=(q_0,0,0,-P_{A,z}))$. For the case of the lepton pair production in the hadron-nucleus collisions we shall consider the centre-of-mass frame of the colliding particles in the high energy limit.

Expressions (5.10), (5.29) and (5.44) for the cross sections show that the structure functions have the following dimensions:

$$[\tilde{W}_1^{lA}]=[m^{-2}],\quad [\tilde{W}_2^{lA}]=[m^{-4}] \tag{5.45a}$$

$$[\tilde{W}_1^{\nu,\bar{\nu}}]=[m^{-2}],\quad [\tilde{W}_2^{\nu,\bar{\nu}}]=[\tilde{W}_3^{\nu,\bar{\nu}}]=[m^{-4}] \tag{5.45b}$$

$$[\rho_1]=[m^{-4}],\quad [\rho_i]=[m^{-6}]\quad i=2,3,\ldots,10 \tag{5.45c}$$

Let's define the Bjorken limit by the following values of the kinematic invariants $s_{lA}, q^2, \nu_A >> m_A^2$; q^2/ν_A is finite. The target fragmentation region is defined as $\nu_{sp} \to \infty$ with the finite ν_{sp}/ν_A and x.

We introduce the longitudinal and transverse scales of momenta and formulate the principle of automodelity:

In the Bjorken limit in the target fragmentation region the structure functions $\tilde{W}_i^{lA}$, $\tilde{W}_i^{\nu,\bar{\nu}}$ and ρ_i transform as homogeneous functions of the corresponding dimensions if the momenta undergo the following scale transformations:

$$p_z \to \lambda p_z, \qquad p_\perp \to p_\perp \tag{5.46}$$

For the deep inelastic lepton-nucleus scattering in the centre-of-mass frame of the nucleus A and the virtual intermediate particle we consider the case $P_{A,z} \to \infty$ and $P^{sp}_{A-1,z} \to \infty$. Then $E^{sp}_{A-1} \approx P^{sp}_{A-1,z} + (P^{sp\,2}_{A-1,\perp} + m^2_{sp})/2P^{sp}_{A-1,z}$ and for the kinematic invariants we obtain

$$\nu_{sp} \approx 2P_{A,z}P^{sp}_{A-1,z} \to \infty$$

$$x \approx \left(\frac{P_{A,z}}{2P^{sp}_{A-1,z}}\right)(P^{sp\,2}_{A-1,\perp} + m^2_{sp}) \quad - \text{ finite.}$$

In accordance with the definition, this case belongs to the target fragmentation region. Under the scale transformations (5.46) of the momenta the kinematic invariants of the deep inelastic scattering of electrons (muons) and neutrinos (antineutrinos) on the nucleus transform in the following manner:

$$\nu_A \to \lambda^2 \nu_A, \qquad q^2 \to \lambda^2 q^2, \qquad \nu_{sp} \to \lambda^2 \nu_{sp}, \qquad x \to x \tag{5.47}$$

It follows from the principle of automodelity that in the Bjorken limit in the target fragmentation region the structure functions of the deep inelastic scattering of charged leptons on the nucleus behave as follows:

$$\tilde{W}_1^{1A}(\lambda^2 q^2, \lambda^2 \nu_A, \lambda^2 \nu_{sp}, x) = \lambda^{-2} \tilde{W}_1^{1A}(q^2, \nu_A, \nu_{sp}, x), \tag{5.48a}$$

$$\tilde{W}_2^{1A}(\lambda^2 q^2, \lambda^2 \nu_A, \lambda^2 \nu_{sp}, x) = \lambda^{-4} \tilde{W}_2^{1A}(q^2, \nu_A, \nu_{sp}, x). \tag{5.48b}$$

For the structure functions of the deep inelastic neutrino-nucleus scattering we obtain

$$\tilde{W}_1^{\nu,\bar{\nu}}(\lambda^2 q^2, \lambda^2 \nu_A, \lambda^2 \nu_{sp}, x) = \lambda^{-2} \tilde{W}_1^{\nu,\bar{\nu}}(q^2, \nu_A, \nu_{sp}, x), \tag{5.49a}$$

$$\tilde{W}_2^{\nu,\bar{\nu}}(\lambda^2 q^2, \lambda^2 \nu_A, \lambda^2 \nu_{sp}, x) = \lambda^{-4} \tilde{W}_2^{\nu,\bar{\nu}}(q^2, \nu_A, \nu_{sp}, x), \tag{5.49b}$$

$$\tilde{W}_3^{\nu,\bar{\nu}}(\lambda^2 q^2, \lambda^2 \nu_A, \lambda^2 \nu_{sp}, x) = \lambda^{-4} \tilde{W}_3^{\nu,\bar{\nu}}(q^2, \nu_A, \nu_{sp}, x). \tag{5.49c}$$

It follows from the Eqs. (5.48a,b) that in the Bjorken limit in the target fragmentation region the structure functions of the deep inelastic scattering of electrons (muons) on the nucleus behave as follows:

$$\lim m_A^2 \tilde{W}_1^{1A}(q^2, \nu_A, \nu_{sp}, x) = f_1^{1A}(q^2/\nu_A, \nu_{sp}/\nu_A, x), \tag{5.50a}$$

$$\lim m_A^2 \nu_A \tilde{W}_2^{1A}(q^2, \nu_A, \nu_{sp}, x) = f_2^{1A}(q^2/\nu_A, \nu_{sp}/\nu_A, x). \tag{5.50b}$$

For the structure functions $\tilde{W}_i^{\nu,\bar{\nu}}$ from Eqs. (5.49a,b,c) we obtain

$$\lim m_A^2 \tilde{W}_1^{\nu,\bar{\nu}}(q^2, \nu_A, \nu_{sp}, x) = f_1^{\nu,\bar{\nu}}(q^2/\nu_A, \nu_{sp}/\nu_A, x), \tag{5.51a}$$

$$\lim m_A^2 \nu_A \tilde{W}_2^{\nu,\bar{\nu}}(q^2, \nu_A, \nu_{sp}, x) = f_2^{\nu,\bar{\nu}}(q^2/\nu_A, \nu_{sp}/\nu_A, x), \tag{5.51b}$$

$$\lim m_A^2 \nu_A \tilde{W}_3^{\nu,\bar{\nu}}(q^2, \nu_A, \nu_{sp}, x) = f_3^{\nu,\bar{\nu}}(q^2/\nu_A, \nu_{sp}/\nu_A, x). \tag{5.51c}$$

The ratio q^2/ν_A is proportional to the well-known Bjorken variable $x_A = -q^2/2\nu_A$. For the ratio ν_{sp}/ν_A, we have $\nu_{sp}/\nu_A \approx x_A^F(1 - x_A^B)$, where the following notations are used:

$$x_A^F = 2P_{A-1,z}^{sp}/\sqrt{s_{qA}}, \qquad s_{qA} = (q + P_A)^2.$$

The invariant x can be expressed in the form $x \approx (P_{A-1,\perp}^{sp\,2} + m_{sp}^2)/2x_A^F$. Taking into account the above remarks, the structure functions of the deep inelastic electron (muon)-nucleus scattering in the Bjorken limit in the target fragmentation region can be written as functions of the x_A^B, x_A^F and $P_{A-1,\perp}^{sp}$:

$$\lim m_A^2 \widetilde{W}_1^{lA}(q^2, \nu_A, \nu_{sp}, x) = F_1^{lA}(x_A^B, x_A^F, P_{A-1,\perp}^{sp}), \tag{5.52a}$$

$$\lim m_A^2 \nu_A \widetilde{W}_2^{lA}(q^2, \nu_A, \nu_{sp}, x) = F_2^{lA}(x_A^B, x_A^F, P_{A-1,\perp}^{sp}). \tag{5.52b}$$

Similarly, for the structure functions of the deep inelastic neutrino-nucleus scattering we obtain

$$\lim m_A^2 \widetilde{W}_1^{\nu,\bar{\nu}}(q^2, \nu_A, \nu_{sp}, x) = F_1^{\nu,\bar{\nu}}(x_A^B, x_A^F, P_{A-1,\perp}^{sp}), \tag{5.53a}$$

$$\lim m_A^2 \nu_A \widetilde{W}_2^{\nu,\bar{\nu}}(q^2, \nu_A, \nu_{sp}, x) = F_2^{\nu,\bar{\nu}}(x_A^B, x_A^F, P_{A-1,\perp}^{sp}), \tag{5.53b}$$

$$\lim m_A^2 \nu_A \widetilde{W}_3^{\nu,\bar{\nu}}(q^2, \nu_A, \nu_{sp}, x) = F_3^{\nu,\bar{\nu}}(x_A^B, x_A^F, P_{A-1,\perp}^{sp}). \tag{5.53c}$$

These results agree with the results obtained in the framework of the Regge-type analysis [151].

We introduce the variable $y = 2\nu_A/s_{lA}$ and rewrite the cross sectios (5.10) and (5.29) as

$$x_A^F \frac{d\sigma(lA \to l'(A-1)X)}{dx_A^B\, dy\, dx_A^F\, d(P_{A-1,\perp}^{sp\,2})} = \frac{1}{8(2\pi)^5}\left(\frac{s_{lA}}{m_A^2}\right)\left(\frac{4\pi\alpha}{q^2}\right)^2 \times$$
$$\times \left\{x_A^B y^2 F_1^{lA}(x_A^B, x_A^F, P_{A-1,\perp}^{sp}) + (1-y) F_2^{lA}(x_A^B, x_A^F, P_{A-1,\perp}^{sp})\right\}, \tag{5.54a}$$

$$x_A^F \frac{d\sigma(\nu A \to l(A-1)X)}{dx_A^B\, dy\, dx_A^F\, d(P_{A-1,\perp}^{sp\,2})} = \frac{G^2}{4(2\pi)^5}\left(\frac{s_{lA}}{m_A^2}\right) \times$$
$$\times \left\{x_A^B y^2 F_1^{\nu,\bar{\nu}}(x_A^B, x_A^F, P_{A-1,\perp}^{sp}) + (1-y) F_2^{\nu,\bar{\nu}}(x_A^B, x_A^F, P_{A-1,\perp}^{sp}) \mp \right. \tag{5.54b}$$
$$\left. \mp x_A^B y\left(1-\frac{y}{2}\right) F_3^{\nu,\bar{\nu}}(x_A^B, x_A^F, P_{A-1,\perp}^{sp})\right\}$$

Integrating these cross sections over the variable y in the from 0 to 1, we obtain

$$x_A^F \frac{d\sigma(lA \to l'(A-1)X)}{dx_A^B\, dx_A^F d(P_{A-1,\perp}^{sp\,2})} = \frac{1}{8(2\pi)^5}\left(\frac{s_{lA}}{m_A^2}\right)\left(\frac{4\pi\alpha}{q^2}\right)^2 \times$$
$$\times \left\{\frac{1}{3} x_A^B F_1^{lA}(x_A^B, x_A^F, P_{A-1,\perp}^{sp}) + \frac{1}{2} F_2^{lA}(x_A^B, x_A^F, P_{A-1,\perp}^{sp})\right\} \quad (5.55a)$$

$$x_A^F \frac{d\sigma(\nu A \to l(A-1)X)}{dx_A^B dx_A^F d(P_{A-1,\perp}^{sp\,2})} = \frac{G^2}{4(2\pi)^5}\left(\frac{s_{lA}}{m_A^2}\right) \times$$
$$\times \left\{\frac{1}{3} x_A^B F_1^{\nu,\bar\nu}(x_A^B, x_A^F, P_{A-1,\perp}^{sp}) + \frac{1}{2} F_2^{\nu,\bar\nu}(x_A^B, x_A^F, P_{A-1,\perp}^{sp}) \mp \right. \quad (5.55b)$$
$$\left. \mp \frac{1}{3} x_A^B F_3^{\nu,\nu}(x_A^B, x_A^F, P_{A-1,\perp}^{sp})\right\}$$

For the process of lepton pair production in hadron-nucleus collisions in the centre-of-mass frame of the colliding particles ($P_A = (E_A, 0, 0, P_{A,z})$, $p_a = (E_a, 0, 0, -P_{A,z})$) we consider the case when $P_{A,z} \to \infty$ and $P_{A-1,z}^{sp} \to \infty$. Then $E_{A-1}^{sp} \approx P_{A-1,z}^{sp} + (P_{A-1,\perp}^{sp} + m_{sp}^2)/2P_{A-1,z}^{sp}$.

As has been already mentioned, not all the arguments of the structure functions $\rho_i(s_{aA}, q^2, \nu_A, \nu_a, \nu_{sp}, x_A, x_a)$ are independent when the transverse part of the four-vector q vanishes and the functions ρ_i can be assumed to be independent of the invariant ν_{sp}. The kinematic invariants of the process are transformed in the following manner

$$s_{aA} \to \lambda^2 s_{aA}, \qquad q^2 \to \lambda^2 q^2, \qquad \nu_A \to \lambda^2 \nu_A,$$
$$\nu_a \to \lambda^2 \nu_a, \qquad x_A \to x_A, \qquad x_a \to \lambda^2 x_a. \quad (5.56)$$

if the momenta undergo the scale transformations (5.46).

According to the principle of automodelity, the functions are transformed as

$$\rho_1(\lambda^2 s_{aA}, \lambda^2 q^2, \lambda^2 \nu_A, \lambda^2 \nu_a, x_A, \lambda^2 x_a) = \lambda^{-4} \rho_1(s_{aA}, q^2, \nu_A, \nu_a, x_A, x_a) \quad (5.57a)$$

$$\rho_i(\lambda^2 s_{aA}, \lambda^2 q^2, \lambda^2 \nu_A, \lambda^2 \nu_a, x_A, \lambda^2 x_a) = \lambda^{-6} \rho_i(s_{aA}, q^2, \nu_A, \nu_a, x_A, x_a) \quad (5.57b)$$
$$i = 2, 3, \ldots, 10$$

Hence, it follows that at large values of the kinematic invariants the structure functions multiplied by some combinations of the kinematic invariants behave as functions of ratios of the kinematic invariants of the process considered. We consider now the two regions:

$$\text{I} \qquad q_z \to \infty \qquad q_0 \approx q_z + q^2/2q_z \,.$$

$$\text{II} \qquad q_z \to -\infty \qquad q_0 \approx |q_z| + q^2/2|q_z| \,.$$

In the region I we construct the arguments of the structure functions ρ_i in the following manner:

$$\begin{aligned}
&\lim m_A^2 q^2 \rho_1(s_{aA}, q^2, \nu_A, \nu_a, x_A, x_a) = \\
&= f_1^{I}(q^2/\nu_A, q^2/\nu_a, \nu_a/s_{aA}, x_a/s_{aA}, x_A), \\
&\lim m_A^4 q^2 \rho_i(s_{aA}, q^2, \nu_A, \nu_a, x_A, x_a) = \\
&= f_i^{I}(q^2/\nu_A, q^2/\nu_a, \nu_a/s_{aA}, x_a/s_{aA}, x_A), \\
&\qquad i = 2, 3, \ldots, 10 .
\end{aligned} \tag{5.58}$$

In region II we choose the corresponding expressions as

$$\begin{aligned}
&\lim m_A^2 q^2 \rho_1(s_{aA}, q^2, \nu_A, \nu_a, x_A, x_a) = \\
&= f_1^{II}(q^2/\nu_A, q^2/\nu_a, \nu_A/s_{aA}, x_a/s_{aA}, x_A) \\
&\lim m_A^4 q^2 \rho_i(s_{aA}, q^2, \nu_A, \nu_a, x_A, x_a) = \\
&= f_i^{II}(q^2/\nu_A, q^2/\nu_a, \nu_A/s_{aA}, x_a/s_{aA}, x_A) \\
&\qquad i = 2, 3, \ldots, 10 .
\end{aligned} \tag{5.59}$$

Let us define the scale invariant variables

$$\begin{aligned}
&x_A^B = -q^2/2\nu_A, \qquad x_a^B = -q^2/2\nu_a, \\
&x_q^F = 2q_z/\sqrt{s_{aA}}, \qquad x_{sp}^F = 2P_{A-1,z}^{sp}/\sqrt{s_{aA}} .
\end{aligned} \tag{5.60}$$

At high energies and in the limit $P^{sp}_{A-1,z} \to \infty$, $q_z \to \pm\infty$ the ratios of the kinematic invariants in Eqs. (5.58) and (5.59) can be expressed in terms of the variables (5.60). Finally, at large values of the kinematic invariants the structure functions of the lepton pair production in hadron-nucleus collisions have the following automodel asymptotics:

$$\lim m_A^2 q^2 \rho_1(s_{aA}, q^2, \nu_A, \nu_a, x_A, x_a) = \\ = F_1(x_A^B, x_a^B, x_q^F, x_{sp}^F, P^{sp}_{A-1,\perp}) \tag{5.61a}$$

$$\lim m_A^4 q^2 \rho_l(s_{aA}, q^2, \nu_A, \nu_a, x_A, x_a) = \\ = F_l(x_A^B, x_a^B, x_q^F, x_{sp}^F, P^{sp}_{A-1,\perp}) \tag{5.61b}$$

$$l = 2, 3, \ldots, 10.$$

5. Relations Between the Nucleus and Nucleon Structure Functions

In this paragraph we derive the relations of the structure functions of the deep inelastic lepton-nucleus scattering and lepton pair production in hadron-nucleus collision to the usual structure functions of the deep inelastic lepton-nucleon scattering and lepton pair production in hadron-nucleon interaction. We shall assume that in the case of lepton-nucleus scattering the incident lepton interacts with only one nucleon of the nucleus A, while the remaining (A-1) nucleons continuing their existence in the form of a fragment nucleus (spectator). Here we restrict our consideration with the case, when virtual photons interact only with nucleons (three-quark states) of nucleus A. More realistic consideration needs taking into acoount of the scattering on multiquark states in nucleus.

We describe the initial nucleus A and the spectator fragment (A-1) in terms of many-body light front relativistic wave functions $\Phi([x_i, \vec{p}_\perp^{(i)}])$.

In this approach the cross section of the process $lA \to l'(A-1)X$ takes the form

$$\frac{d\sigma(lA \to l'(A-1)X)}{dq^2 d\nu_A d\nu_{sp} dx} = \frac{\lambda(s_{lN}, m_l^2, m_N^2)}{2(2\pi)^2 \lambda(s_{lA}, m_l^2, m_A^2)} \times$$
$$\times (\nu_A^2 - m_A^2 q^2)^{-1/2} \left| \frac{I(X^{sp}, \vec{P}^{sp}_{A-1,\perp})}{1 - X^{sp}} \right|^2 \frac{d\sigma(lN \to l'X)}{dq^2 d\nu_N} \tag{5.62}$$

Here $d\sigma(lN \to l'X)/dq^2 d\nu_N$ is the cross section of the deep inelastic lepton-nucleon interaction, m_N is the nucleon mass. The kinematic invariants

$$s_{lN} = (k + p_N)^2, \qquad \nu_N = p_N q \tag{5.63}$$

of the lepton-nucleon interaction are related to the invariants of the lepton-nucleus interaction as follows:

$$s_{lN} = s_{lA} + m_{sp}^2 - 2kP^{sp}_{A-1} - 2x, \qquad \nu_N = \nu_A - \nu_{sp} \tag{5.64}$$

where p_N is the four-momentum of the nucleon from the nucleus A that interacts with the lepton, $I(X^{sp}, \vec{P}^{sp}_{A-1,\perp})$ is the overlap integral of the wave functions of the initial nucleus and the fragment nucleus :

$$I(X^{sp}, \vec{P}^{sp}_{A-1,\perp}) = \int_0^1 \prod_{i=1}^{A-1} \frac{dx_i^{(A-1)'}}{x_i^{(A-1)'}} \delta(1 - \sum_{i=1}^{A-1} x_i^{(A-1)'}) \times$$
$$\times \int \prod_{i=1}^{A-1} d\vec{p}_\perp^{(i)'} \delta(\vec{P}^{sp}_{A-1,\perp} - \sum_{i=1}^{A-1} \vec{p}_\perp^{(i)'}) \times \tag{5.65}$$
$$\times \Phi^+_{P^{sp}_{A-1}}([x_i^{(A-1)'}, \vec{p}_\perp^{(i)'}]) \Phi_{P_A}([x_i^{(A)}, \vec{p}_\perp^{(i)}])$$

The variable X^{sp} is defined in the following way

$$X^{sp}=(P^{sp}_{A-1,0}+P^{sp}_{A-1,z})/(P_{A,0}+P_{A,z})$$

The variables $x_i^{(A)}$ and $\vec{p}_\perp^{\,(i)}$ are related to the integration variables $x_i^{(A-1)'}$ and $\vec{p}_\perp^{\,(i)'}$ in the following simple way:

$$x_i^{(A)}=X^{sp}x_i^{(A-1)'},\quad \vec{p}_\perp^{\,(i)}=\vec{p}_\perp^{\,(i)'},\quad i=1,2,\dots,A-1$$

$$x_A^{(A)}=1-X^{sp},\quad \vec{p}_\perp^{\,(A)}=-\vec{P}^{sp}_{A-1,\perp}\quad \text{in the frame, where } \vec{P}_{A,\perp}=0.$$

In the case of inclusive photoabsorption on the nucleus the following expression for the cross section can be obtained:

$$\left\langle\frac{d\sigma(\gamma^*A\to(A-1)X)}{d\vec{P}^{sp}_{A-1}/E^{sp}_{A-1}}\right\rangle=\frac{1}{2(2\pi)^2}\frac{(\nu_N+q^2/2)}{(\nu_A+q^2/2)}\times$$
$$\times\left|\frac{I(X^{sp},\vec{P}^{sp}_{A-1,\perp})}{1-X^{sp}}\right|^2\sigma(\gamma^*N\to X) \tag{5.66}$$

Here $\sigma(\gamma^*N\to X)$ is the total photoabsorption cross section on the nucleon. Using the Eqs. (5.10), (5.14), (5.62), (5.66) and the well-known expressions for the cross sections of deep inelastic electroproduction and the nucleon-photoabsorption (see, e.g., [152]), we obtain the relations between the structure functions $\tilde{W}_i^{IA}(q^2,\nu_A,\nu_{sp},x)$ and the structure functions $W_i^{IN}(q^2,\nu_N)$:

$$\tilde{W}_1^{IA}(q^2,\nu_A,\nu_{sp},x)=\frac{1}{2(2\pi)^2}\left|\frac{I(X^{sp},\vec{P}^{sp}_{A-1,\perp})}{1-X^{sp}}\right|^2W_1^{IN}(q^2,\nu_N)$$

$$\tilde{W}_2^{IA}(q^2,\nu_A,\nu_{sp},x)=\frac{1}{2(2\pi)^2}\left|\frac{I(X^{sp},\vec{P}^{sp}_{A-1,\perp})}{1-X^{sp}}\right|^2\times \tag{5.67}$$
$$\times\frac{(\nu_N^2-m_N^2q^2)}{(\nu_A^2-m_A^2q^2)}W_2^{IN}(q^2,\nu_N)$$

The variables X^{sp} and $P^{sp}_{A-1,\perp}$ can be expressed by means of the invariants q^2, ν_A, ν_{sp} and x:

$$X^{sp} = (1/m_A^2)[x + (x\nu_A - \nu_{sp} m_A^2)/(\nu_A^2 - m_A^2 q^2)^{1/2}]$$

$$P^{sp\,2}_{A-1,\perp} = (1/q^2)[\nu_{sp}^2 - (\nu_A \nu_{sp} - q^2 x)^2/(\nu_A^2 - m_A^2 q^2)] - m_{sp}^2$$

Taking into account the automodel behaviour of the structure functions $W_1^{IN}(q^2, \nu_N)$ and $\nu_N W_2^{IN}(q^2, \nu_N)$ in the Bjorken limit, we obtain the following properties of scale invariance for the structure functions $\tilde{W}_i^{lA}$ in the target fragmentation region:

$$\lim \tilde{W}_1^{lA}(q^2, \nu_A, \nu_{sp}, x) = \frac{1}{2(2\pi)^2} \left| \frac{I(x_A^F, \vec{P}^{sp}_{A-1,\perp})}{1 - x_A^F} \right|^2 \times \\ \times \varphi_1^{IN}\left(\frac{x_A^B}{1 - x_A^F(1 - x_A^B)}\right) \tag{5.68a}$$

$$\lim \nu_A \tilde{W}_2^{lA}(q^2, \nu_A, \nu_{sp}, x) = \frac{1}{2(2\pi)^2} \left| \frac{I(x_A^F, \vec{P}^{sp}_{A-1,\perp})}{1 - x_A^F} \right|^2 \times \\ \times [1 - x_A^F(1 - x_A^B)]\, \varphi_2^{IN}\left(\frac{x_A^B}{1 - x_A^F(1 - x_A^B)}\right) \tag{5.68b}$$

A relation between the cross sections of the deep inelastic neutrino (anti-neutrino)-nucleus scattering and the neutrino-nucleon scattering can also be obtained. It is of the form

$$\frac{d\sigma(\nu A \to l(A-1)X)}{dq^2 d\nu_A d\nu_{sp} dx} = \frac{(s_{\nu N} - m_N^2)^2}{2(2\pi)^2 (s_{\nu A} - m_A^2)^2} (\nu_A^2 - m_A^2 q^2)^{-1/2} \times \\ \times \left| \frac{I(x_A^F, \vec{P}^{sp}_{A-1,\perp})}{1 - X^{sp}} \right|^2 \frac{d\sigma(\nu N \to lX)}{dq^2 d\nu_N} \tag{5.69}$$

The cross section of the absorption of the virtual massless particle with quantum numbers of the $W^{\pm}$-boson on the nucleus and on the nucleon are related to each other as follows:

$$\left\langle \frac{d\sigma(WA\to(A-1)X)}{d\vec{P}^{sp}_{A-1}/E^{sp}_{A-1}} \right\rangle = \frac{1}{2(2\pi)^2}\,\frac{(\nu_N+q^2/2)}{(\nu_A+q^2/2)}\times \tag{5.70}$$

$$\times\left|\frac{I(X^{sp},\vec{P}^{sp}_{A-1,\perp})}{1-X^{sp}}\right|^2 \sigma(WN\to X)$$

Here $\sigma(WN\to X)$ is the total cross section of absorption of W-boson on nucleon. Using the Eqs. (5.29), (5.32), (5.69), (5.70) and well-known expressions for cross sections of the deep inelastic neutrino-nucleon interactions (see, e.g., [152]), we obtain the following relations between the structure functions $\tilde{W}_i^{\nu,\bar{\nu}}(q^2,\nu_A,\nu_{sp},x)$ and $W_i^{\nu,\bar{\nu}}(q^2,\nu_N)$:

$$\tilde{W}_1^{\nu,\bar{\nu}}(q^2,\nu_A,\nu_{sp},x)=(4\pi)^2\left|\frac{I(X^{sp},\vec{P}^{sp}_{A-1,\perp})}{1-X^{sp}}\right|^2 W_1^{\nu,\bar{\nu}}(q^2,\nu_N) \tag{5.71a}$$

$$\tilde{W}_2^{\nu,\bar{\nu}}(q^2,\nu_A,\nu_{sp},x)=(4\pi)^2\left|\frac{I(X^{sp},\vec{P}^{sp}_{A-1,\perp})}{1-X^{sp}}\right|^2\times \tag{5.71b}$$

$$\times\frac{(\nu_N^2-m_N^2q^2)}{(\nu_A^2-m_A^2q^2)}W_2^{\nu,\bar{\nu}}(q^2,\nu_N)$$

$$\tilde{W}_3^{\nu,\bar{\nu}}(q^2,\nu_A,\nu_{sp},x)=(4\pi)^2\left|\frac{I(X^{sp},\vec{P}^{sp}_{A-1,\perp})}{1-X^{sp}}\right|^2\times \tag{5.71c}$$

$$\times\frac{(\nu_N^2-m_N^2q^2)^{1/2}}{(\nu_A^2-m_A^2q^2)^{1/2}}W_3^{\nu,\bar{\nu}}(q^2,\nu_N)$$

Taking into account the automodel properties of the structure functions $W_1^{\nu,\bar{\nu}}$, $\nu_N W_2^{\nu,\bar{\nu}}$ and $\nu_N W_3^{\nu,\bar{\nu}}$ of the deep inelastic neutrino-nucleon interaction in the Bjorken limit, we obtain the following properties of scale invariance for the structure functions $\tilde{W}_i^{\nu,\bar{\nu}}(q^2,\nu_A,\nu_{sp},x)$ in the target fragmentation region:

$$\lim \tilde{W}_1^{\nu,\bar{\nu}}(q^2,\nu_A,\nu_{sp},x) = (4\pi)^2 \left|\frac{I(x_A^F,\vec{P}^{sp}_{A-1,\perp})}{1-x_A^F}\right|^2 \times$$
$$\times\, \varphi_1^{\nu,\bar{\nu}}\left(\frac{x_A^B}{1-x_A^F(1-x_A^B)}\right) \tag{5.72a}$$

$$\lim \nu_A \tilde{W}_2^{\nu,\bar{\nu}}(q^2,\nu_A,\nu_{sp},x) = (4\pi)^2 \left|\frac{I(x_A^F,\vec{P}^{sp}_{A-1,\perp})}{1-x_A^F}\right|^2 \times$$
$$\times\, [1-x_A^F(1-x_A^B)]\, \varphi_2^{\nu,\bar{\nu}}\left(\frac{x_A^B}{1-x_A^F(1-x_A^B)}\right) \tag{5.72b}$$

$$\lim \nu_A \tilde{W}_3^{\nu,\bar{\nu}}(q^2,\nu_A,\nu_{sp},x) = (4\pi)^2 \left|\frac{I(x_A^F,\vec{P}^{sp}_{A-1,\perp})}{1-x_A^F}\right|^2 \times$$
$$\times\, \varphi_3^{\nu,\bar{\nu}}\left(\frac{x_A^B}{1-x_A^F(1-x_A^B)}\right) \tag{5.72c}$$

Note that the automodel properties of the structure functions $\tilde{W}_i^{lA}$ and $\tilde{W}_i^{\nu,\bar{\nu}}$ are in agreement with the results of the principle of automodelity.

In Eqs. (5.67) and (5.71) the structure functions W_i^{lN} and $W_i^{\nu,\bar{\nu}}$ of the nucleon should be taken, in general, off mass shell, since corresponding nucleon is bound in the nucleus. While there is no reliable information for these quantities it is assumed that the functions W_i^{lN} and $W_i^{\nu,\bar{\nu}}$ off mass shell have the same analytic form as on the mass shell, but the arguments take values off this surface.

We consider now the lepton pair production process in hadron-nucleus collision. We assume that production of lepton pair takes place in interaction of incident hadron with one nucleon of the nucleus A. In this case it is possible to derive relation between the structure functions of lepton pair production processes in hadron-nucleus and hadron-nucleon collisions. In this approximation the cross sections of the processes $aA \to l^+l^-(A-1)X$ and

$aN \to l^+l^- X$ are related in the following manner:

$$\frac{d\sigma(aA \to l^+l^-(A-1)X)}{dq^2 d\nu_A d\nu_a dx_A dx_a} = \frac{\lambda(s_{aA}, m_a^2, m_N^2)}{(2\pi)^2(4\pi)^{2A-1}\lambda^{3/2}(s_{aA}, m_a^2, m_A^2)} \times$$

$$\times \left|\frac{I(X^{sp}, \vec{P}^{sp}_{A-1,\perp})}{1-X^{sp}}\right|^2 \frac{d\sigma(aN \to l^+l^-X)}{dq^2 d\nu_A d\nu_a} \qquad (5.73)$$

The kinematic invariants of hadron-nucleon collision

$$s_{aN} = (p_a + p_N)^2, \qquad \nu_N = p_N q$$

are rlated with the invariants of hadron-nucleus collision in the following way

$$s_{aN} = s_{aA} - 2x_A - 2x_a + m_{sp}^2, \qquad \nu_N = \nu_A - \nu_{sp} \qquad (5.74)$$

Using the Eqs. (5.44), (5.73) and the well-known expression for the cross section of lepton pair production in hadron-nucleon collision (see, e.g., [152]), we can obtain the following relation between the structure functions $\tilde{\rho}_i^{aA}$ and ρ_i^{aN}:

$$\tilde{\rho}_1^{aA} = \frac{2\pi}{(4\pi)^{2A-2}} \left|\frac{I(X^{sp}, \vec{P}^{sp}_{A-1,\perp})}{1-X^{sp}}\right|^2 \rho_1^{aN} \qquad (5.75a)$$

$$\tilde{\rho}_2^{aA} = \frac{2\pi}{(4\pi)^{2A-2}} \left|\frac{I(X^{sp}, \vec{P}^{sp}_{A-1,\perp})}{1-X^{sp}}\right|^2 \frac{(\nu_N^2 - m_N^2 q^2)}{(\nu_A^2 - m_A^2 q^2)} \rho_2^{aN} \qquad (5.75b)$$

$$\tilde{\rho}_3^{aA} = \frac{2\pi}{(4\pi)^{2A-2}} \left|\frac{I(X^{sp}, \vec{P}^{sp}_{A-1,\perp})}{1-X^{sp}}\right|^2 \rho_3^{aN} \qquad (5.75c)$$

$$\tilde{\rho}_4^{aA} = \frac{2\pi}{(4\pi)^{2A-2}} \left|\frac{I(X^{sp}, \vec{P}^{sp}_{A-1,\perp})}{1-X^{sp}}\right|^2 \frac{(s_{aN} - m_N^2 - m_a^2 - 2\nu_N\nu_a/q^2)}{(s_{aA} - m_A^2 - m_a^2 - 2\nu_A\nu_a/q^2)} \rho_4^{aN} \qquad (5.75d)$$

Here the structure functions $\tilde{\rho}_i^{aA}$ are functions of the kinematic invariants

s_{aA}, q^2, ν_A, ν_a, x_A, x_a and the structure functions ρ_i^{aN} are functions of s_{aN}, q^2, ν_A, ν_a. Assuming that at large values of their arguments the structure functions ρ_i^{aN} are functions of ratios of these invariants (i.e. they have automodel behaviour), we obtain that in the high energy limit the functions $\tilde{\rho}_i^{aA}$ are functions of scale-invariant variables (5.60) in agreement with the results of the principle of automodelity.

The expressions obtained in this section make it possible to study the structure of the nucleus in a relativistic manner. For example, by measuring experimentally the structure functions $\tilde{W}_i^{lA}$ and $\tilde{W}_i^{\nu,\bar{\nu}}$ and using information on the structure functions W_i^{lN} and $W_i^{\nu,\bar{\nu}}$ (detailed information about the fstructure functions of the deep inelastic lepton-nucleon scattering can be found, e.g., in [153]), one can study the overlap integral $I(\chi^{sp}, \vec{p}^{\,sp}_{A-1,\perp})$, which carries information about the relativistic wave functions of the initial and final nuclei.

R E F E R E N C E S

1. A.M.Baldin, S.B.Gerasimov et al., in: Lectures at the School on High Energy Physics, Sukhumi, 1972. R2-6867, JINR, Dubna, 1972, p. 374.
2. V.P.Alekseev, A.M.Baldin et al., Preprint JINR 9-7148, Dubna, 1973.
3. H.Steiner, Talk presented at the Intern. Conf. on High Energy Collisions Involving Nuclei, Trieste, Italy, 1974. Preprint LBL-3613, Berkeley, 1974.
4. A.M.Baldin, Talk presented at the 6th Intern. Conf. on High Energy Physics and Nuclear Structure, Santa-Fe, USA, 1975. Preprint JINR E2-9138, Dubna, 1975.
5. G.A.Leksin, Lectures at the School on Some Problems of Experimental Investigations in High Energy Physics, Moscow Engineering Physics Institute, 1975.
6. P.Zielinski, in: Proc. of the 18th Intern. Conf. on High Energy Physics, Tbilisi, USSR, 1976. JINR D1,2-10400, Dubna, 1976.
7. V.R.Garsevanishvili, Lectures at the 13th Intern. Winter School of Theoretical Physics, Karpacz, Poland, 1976.
8. A.M.Baldin, Sov. J. Part. Nucl., 8, 175 (1977).
9. H.Steiner, in: Proc. of the 7th Intern. Conf. on High Energy Physics and Nuclear Structure, Zurich, Switzerland, 1977. Ed. M.Locher. 1977.
10. A.M.Baldin, in: Proc. of the 7th Intern. Conf. on High Energy Physics and Nuclear Structure, Zurich, Switzerland, 1977. Ed. M.Locher. 1977.
11. A.S.Goldhaber, H.H.Heckman, Ann. Rev. Nucl. Part. Sci., 28, 161 (1978).
12. V.A.Karmanov, I.S.Shapiro, Sov. J. Part. Nucl., 9, 134 (1978).
13. A.M.Baldin, in: Fundamental Problems of Theoretical and Mathematical Physics. JINR D-12831, Dubna, 1979.
14. V.K.Luk'yanov, A.I.Titov, Sov. J. Part. Nucl., 10, 321 (1979).
15. V.S.Stavinskii, Sov. J. Part. Nucl., 10, 373 (1979).
16. A.M.Baldin, Talk at the Intern. Conf. on Extreme State in Nuclear Systems, Dresden, GDR, 1980. Preprint JINR E1-80-174, Dubna, 1980.
17. M.I.Strikman, L.L.Frankfurt, Sov. J. Part. Nucl., 11, 221 (1980); Phys. Rep., C76, 215 (1981).

18. A.M.Baldin, in: Proc. of the 1981 CERN-JINR School of Physics, Hanko, Finland, 1981. CERN 82-04, Geneve, 1982.

19. E.M.Friedlander, H.H.Heckman, Preprint LBL-13864, Berkeley, 1982.

20. A.V.Efremov, Sov. J. Part. Nucl., 13, 254 (1982).

21. V.K.Luk'yanov, A.I.Titov, in: Proc. of the 15th Intern. School on High Energy Physics, Dubna, USSR, 1982. JINR D2,4-83-179, Dubna, 1983.

22. J.Cleymans, M.Dechantreiter, F.Halzen, Univ. of Wisconsin preprint MAD/TH/50, Madison, 1982.

23. A.P.Kobushkin, V.P.Shelest, Sov. J. Part. Nucl., 14, 483 (1989).

24. L.Van Hove, Lecture at the Intern. Conf. on Nuclear Physics, Florence, Italy, 1983. Preprint CERN TH. 3725, Geneva, 1983.

25. V.R.Garsevanishvili, Z.R.Menteshashvili, D.G.Mirianashvili, M.S.Nioradze, Sov. J. Part. Nucl., 15, 497 (1984).
V.R.Garsevanishvili, in: Proc. of the 9th European Conf. on Few Body Problems in Physics, Tbilisi, Georgia, USSR, 1984. Eds. L.D.Faddeev and T.I.Kopaleishvili, World Scientific, Singapore, 1984.

26. M.A.Faessler, Phys. Rep., 115, 1 (1984).

27. M.I.Strikman, L.L.Frankfurt, Phys. Rep., 160, 235 (1988).

28. L.Kluberg, in: Proc. of 24th Intern. Conf. on High Energy Physics, Munich, 1988. Eds. R.Kotthaus and J.H.Kuhn, Springer-Verlag, 1988, p. 250.

29. V.A.Karmanov, Sov. J. Part. Nucl., 19, 525 (1988).

30. G.C.Mantovani et al., Phys. Lett., 64B, 471 (1976).

31. G.C.Mantovani et al., Phys. Lett., 65B, 401 (1976).

32. G.Alberi, G.Goggi, Phys. Rep., 74, 1 (1981).

33. G.Jakob, Th.Maris, Rev. Mod. Phys., 45, 6 (1973).

34. C.Ciofi degli Atti, in: Photonuclear Reactions. Eds. S.Costa and C.Schaerf, Springer-Verlag, Berlin, 1977, p. 521.

35. A.M.Baldin, Short Communications in Physics, 1, 35 (1971).

36. A.M.Baldin et al., Preprint JINR P1-5819, Dubna, 1971; Sov. J. Nucl. Phys., 18, 41 (1973); 21, 517 (1975).

37. E.E.Salpeter, H.A.Bethe, Phys. Rev., 84, 1232 (1951).
M.Gell-Mann, F.Low, Phys. Rev., 84, 350 (1951).

38. A.A.Logunov, A.N.Tavkhelidze, Nuovo Cim., 29, 380 (1963).
V.G.Kadyshevsky, A.N.Tavkhelidze, in: Problems of Theoretical Physics, dedicated to N.N.Bogolubov in the occasion of his 60th birthday. Nauka, Moscow, 1969.

39. S.J.Brodsky, in: Quarks and Nuclear Forces. Eds. D.C.Fries and B.Zeitnitz, Springer-Verlag, Berlin, Heidelberg, New York, 1982, p. 81.

40. P.A.M.Dirac, Rev. Mod. Phys., 21, 392 (1949).

41. S.-J. Chang, R.G.Root, T.-M.Yan, Phys. Rev., D7, 1133 (1973).

42. S.-J.Chang, T.-M.Yan, Phys. Rev., D7, 1147 (1973).

43. T.-M.Yan, Phys. Rev., D7, 1760 (1973).

44. T.-M.Yan, Phys. Rev., D7, 1780 (1973).

45. N.M.Atakishiev, R.M.Mir-Kasimov, Sh.M.Nagiev, Theor. Mat. Phys., 32, 34 (1977).

46. V.R.Garsevanishvili, A.N.Kvinikhidze, V.A.Matveev, A.N.Tavkhelidze, R.M.Faustov, Theor. Mat. Phys., 23, 310 (1975).

47. A.A.Khelashvili, Preprint JINR P2-8750, Dubna, 1975.

48. L.Susskind, Phys. Rev., 165, 1535 (1968).

49. J.B.Kogut, D.E.Soper, Phys. Rev., D1, 2901 (1970).
J.B.Kogut, L.Susskind, Phys. Rep., 8, 75 (1973).

50. J.F.Gunion, S.J.Brodsky, R.Blankenbecler, Phys. Rev., D8, 287 (1973).

51. S.Weinberg, Phys. Rev., 150, 1313 (1966).

52. V.R.Garsevanishvili, V.A.Matveev, Theor. Mat. Phys., 24, 3 (1975).

53. N.N.Singh, Y.K.Mathur, A.N.Mitra, Few-Body Systems, 1, 47 (1986).

54. D.S.Kulshreshtha, A.N.Mitra, Phys. Rev., D37, 1268 (1988).

55. B.H.J.McKellar, M.D.Scadron, R.C.Warner, Int. J. of Mod. Phys., A3, 203 (1988).

56. H.Leutwyler, Nucl. Phys., 76B, 413 (1974).

57. H.Leutwyler, J.Stern, Ann. Phys., 112, 94 (1978).

58. M.V.Terent'ev, Sov. J. Nucl. Phys., 24, 207 (1976).

59. V.B.Berestetskii, M.V.Terent'ev, Sov. J. Nucl. Phys., 24, 547 (1976).

60. B.L.Bakker, L.A.Kondratyuk, M.V.Terent'ev, Nucl. Phys., B158, 497 (1979).

61. N.N.Bogolubov, D.V.Shirkov, Introduction to the Theory of Quantized Fields, 3rd Ed. Wiley, New York, 1980.

62. S.Mandelstam, Proc. Royal Soc., 233A, 248 (1955).
63. R.N.Faustov, Ann. Phys., 78, 176 (1973).
64. N.N.Bogolubov, V.A.Matveev, A.N.Tavkhelidze, in: Proc. of the Intern. Seminar on Elementary Particle Physics, Varna, Bulgaria, 1968. JINR P2-4050, Dubna, 1968, p. 269.
65. C.Alabiso, G.Schierholz, Phys. Rev., D10, 960 (1974).
66. R.N.Faustov, A.A.Khelashvili, Sov. J. Nucl. Phys., 10, 619 (1970).
67. V.A.Matveev, R.M.Muradyan, A.N.Tavkhelidze, Lett. Nuovo Com., 7, 719, (1973).
68. S.J.Brodsky, G.Farrar, Phys. Rev. Lett., 31, 1153 (1973).
69. V.R.Garsevanishvili, A.N.Kvinikhidze, V.A.Matveev, A.N.Tavkhelidze, R.N.Faustov, Theor. Mat. Phys., 25, 37 (1975).
70. V.A.Matveev, R.M.Muradyan, A.N.Tavkhelidze, Preprint JINR E2-8048, Dubna, 1974.
71. D.Sivers, S.J.Brodsky, R.Blankenbecler, Phys. Rep., 23, 1 (1976).
72. A.N.Kvinikhidze, V.A.Matveev, A.N.Sisakyan, L.A.Slepchenko, A.N.Tavkhelidze, in: Proc. of the Intern. Workshop "Multiple Production Processes and Inclusive Reactions at High Energies", Serpukhov, USSR, 1976. Serpukhov, 1977, p. 300.
73. R.P.Feynman, Phys. Rev. Lett., 23, 1415 (1969).
74. J.D.Bjorken, E.A.Paschos, Phys. Rev., 185, 1975 (1969).
75. J.D.Bjorken, E.A.Paschos, Phys. Rev, D1, 3151 (1970).
76. R.P.Feynman, Photon-Hadron Interactions, W.A.Benjamin, Reading, Mass., 1972.
77. A.Bodek et al., Phys. Rev. Lett., 30, 1087 (1973).
78. E.M.Riordan et al., Phys. Rev. Lett., 33, 561 (1974).
79. E.M.Riordan et al., Preprint SLAC-PUB 1634, Stanford, 1975.
80. A.Bodek et al., Phys. Rev, D20, 1471 (1979).
81. B.A.Gordon et al., Phys. Rev., D20, 2645 (1979).
82. O.Nachtmann, Nucl. Phys., B63, 237 (1973).
83. A.de Rujula, H.Georgi, H.D.Politzer, Phys. Rev., D10, 2141 (1974).
84. H.Georgi, H.D.Politzer, Phys. Rev., D14, 1829 (1976).

85. A.de Rujula, H.Georgi, H.D.Politzer. Ann. Phys., 103, 315 (1977).
86. A.V.Efremov, A.V.Radyushkin, in: Proc. of Intern. Seminar "Multiple Production Processes and Inclusive Reactions at High Energies", Serpukhov, USSR, 1976. Serpukhov, 1977, p. 252.
87. N.N.Krasnikov, A.N.Tavkhelidze, K.G.Chetirkin, in: Proc. of the Intern. Conf. "Neutrino-77". Nauka Publ., Moscow, 1978, v.2, p. 189.
88. N.M.Atakishiev, R.M.Mir-Kasimov, Sh.M.Nagiev, Preprint JINR P2-80-635, Dubna, 1980.
89. R.Arnold et al., Phys. Rev. Lett., 35, 776 (1975); 40, 1429 (1978).
90. J.J.Aubert et al., Phys. Lett., 123B, 275 (1983).
91. A.Bodek et al., Phys. Rev. Lett., 50, 1431 (1983); 51, 534 (1983).
92. R.G.Arnold et al., Phys. Rev. Lett., 52, 727 (1984).
93. G.Bari et al., Phys. Lett., 163B, 282 (1985).
94. C.H.Llewellyn Smith, Phys. Lett., 128B, 107 (1983).
95. M.Ericson, A.W.Thomas, Phys. Lett., 128B, 112 (1983).
96. F.E.Close, R.G.Roberts, G.G.Ross, Phys. Lett., 129B, 346 (1983).
97. C.E.Carlson, J.J.Havens, Phys. Rev. Lett., 51, 261 (1983).
98. S.Date, Progr. Theor. Phys., 70, 1682 (1983).
99. E.L.Berger, F.Coester, R.B.Wiringa, Phys. Rev., D29, 383 (1984).
100. R.L.Jaffe, F.E.Close, R.G.Roberts, G.G.Ross, Phys. Lett., 134B, 449 (1984).
101. O.Nachtman, H.J.Pirner, Z.Phys., C21, 277 (1984).
102. A.I.Titov, Sov. J. Nucl. Phys., 40, 50 (1984).
103. W.Furmanski, A.Krzywicki, Z.Phys., C22, 391 (1984).
104. L.A.Kondratyuk, M.J.Shmatikov, JETP Lett., 39, 389 (1984).
105. J.P.Vary, Nucl. Phys., A418, 195c (1984).
106. S.V.Akulinichev, S.A.Kulagin, G.M.Vagradov, JETP Lett., 42, 127 (1985); Phys. Lett., 152B, 485 (1985).
107. B.L.Birbrair, A.B.Gridnev, M.B.Zhalov, E.M.Levin, V.E.Starodubsky, Phys. Lett., 166B, 119 (1986).
108. V.V.Malyarov, Theory of Atomic Nuclei. Nauka Publ., Moscow, 1967.
109. S.Date, A.Nakamura, Progr. Theor. Phys., 69, 565 (1983).
110. V.A.Matveev, P,Sorba, Lett. Nuovo Cim., 20, 443 (1977).

111. P.J.Mulders, A.W.Thomas, Phys. Rev. Lett., 52, 1199 (1984).
112. A.M.Baldin, Preprint JINR E2-83-415, Dubna, 1983.
113. I.A.Savin, in: Proc. of the 6th Intern. Seminar on High Energy Physics Problems, Dubna, USSR, 1981. JINR D1,2-81-728, Dubna, 1981, p. 223.
114. J. Ashman et al., Phys. Lett., 202B, 603 (1988).
115. M.Arneodo et al., Phys. Lett., 211B, 493 (1988).
116. R.J.M.Covolan, E.Predazzi, Preprint DFTT 21/90, Torino, 1990.
117. V.R.Garsevanishvili, D.G.Mirianashvili, M.S.Nioradze, Communications JINR P2-9859, Dubna, 1976.
118. V.R.Garsevanishvili, D.G.Mirianashvili, Rep. Math. Phys., 11, 89 (1977).
119. I.S.Shapiro, in: Proc. of the Intern. Conf. on Selected Problems of Nuclear Structure, Dubna, 1976. JINR D-9920, Dubna, 1976, p. 424.
120. V.V.Burov, V.K.Luk'yanov, A.I.Titov, in: Proc. of the Intern. Conf. on Selected Problems of Nuclear Structure, Dubna, 1976. JINR D-9920, Dubna, 1976, p. 432.
121. W.Buck, F.Gross, Phys. Lett., 63B, 286 (1976).
122. V.A.Karmanov, Sov. J. JETP, 44, 210 (1976).
123. M.I.Strikman, L.L.Frankfurt, Sov. J. Nucl. Phys., 25, 1177 (1977).
124. I.Schmidt, R.Blankenbecler, Phys. Rev., D15, 3321 (1977).
125. W.Buck, F.Gross, Phys. Rev., D20, 2361 (1979).
126. B.S.Aladashvili, V.R.Garsevanishvili, V.V.Glagolev, Z.R.Menteshashvili et al., Sov. J. Nucl. Phys., 33, 681 (1981).
127. J.Benecke, T.T.Chou, C.N.Yang, E.Yen, Phys. Rev., 188, 2159 (1969).
128. V.A.Matveev, R.M.Muradyan, A.N.Tavkhelidze, Lett., Nuovo Cim., 5, 907 (1972); Theor. Mat. Phys., 15, 332 (1973).
129. V.R.Garsevanishvili, V.V.Glagolev, Z.R.Menteshashvili et al., Preprint JINR 1-81-838, Dubna, 1981.
130. B.S.Aladashvili et al., Nucl. Phys., 86B, 461 (1975); J.Phys. G: Nucl. Phys., 3, 1225 (1976); Sov. J. Nucl. Phys., 27, 377 (1978).
131. A.Fridman. Fortschr. Phys., 23, 243 (1975).
132. O.Benary et al., "A Compilation", UCRL-20000NN, Berkeley, 1970.
V.Flaminio et al., "A Compilation", CERN-HERA 79-03, Geneva, 1979.

133. A.Abdivaliev, K.Beshliu, V.R.Garsevanishvili et al., Sov. J. Nucl. Phys., 27, 715 (1978).

134. A.P.Gasparyan et al., Preprint JINR 1-9111, Dubna, 1975.

135. F.Kotorobai et al., Preprint JINR P10-9314, Dubna, 1975.

136. A.P.Ierusalimov et al., Preprint JINR P10-9502, Dubna, 1976.

137. V.R.Garsevanishvili, Z.R.Menteshashvili, D.G.Mirianashvili, M.S.Nioradze, Theor. Mat. Phys., 33, 276 (1977).

138. B.S.Aladashvili, V.R.Garsevanishvili, Z.R.Menteshashvili et al., Sov. J. Nucl. Phys., 34, 591 (1981).

139. V.R.Garsevanishvili, Z.R.Menteshashvili, D.G.Mirianashvili, M.S.Nioradze, Sov. J. Nucl. Phys., 41, 587 (1985).

140. V.V.Glagolev et al., Phys. Rev., C18, 1382 (1979); Communication JINR E1-12943, Dubna, 1979.

141. N.A.Buzdavina, V.V.Glagolev et al., Communication JINR 1-81-530, Dubna, 1981.

142. G.Bizard et al., Nucl. Phys., A285, 461 (1977).

143. M.V.Fedoryuk, The Method of Steepest Descent, Nauka, Moscow, 1977.

144. M.M.Block. Nuovo Cim., 20, 715 (1961).

145. B.Z.Kopeliovich, I.K.Potashnikova, Sov. J. Nucl. Phys., 13, 592 (1971).

146. E.M.Levin, M.I.Strikman, Sov. J. Nucl. Phys., 23, 216 (1976).

147. J.P.Auger, J.Gillespie, R.J.Lombard, Nucl. Phys., A262, 372 (1976).

148. G.Alberi, L.P.Rosa, Z.D.Thome, Phys. Rev. Lett., 34, 503 (1975).

149. V.R.Garsevanishvili, Ya.Z.Darbaidze, Z.R.Menteshashvili, Theor. Mat. Phys., 42, 50 (1980).

150. Ya.Z.Darbaidze, Sh.M.Esakia, V.R.Garsevanishvili, Z.R.Menteshashvili, Fortschr. Phys., 28, 501 (1980).

151. N.K.Pak, Preprint SLAC-PUB 1744, Stanford, 1976.

152. V.A.Matveev, in: Proc. of the 1973 CERN-JINR School of Physics, Ebeltoft, Denmark, 1973. CERN 73-12, Geneva, 1973, p. 251.

153. M.Klein, W.D.Nowak, Preprint JINR E1-11542, Dubna, 1978.

Subject Index

Automodelity, Dimensional Analysis and the Principle of, 123-128
Bound State Wave Function, Equation for the Many-Body, 15-20
Bound State Wave Function, Equation for the Two-Body, 9-12
Bound States, Light Front Formalism for, 9-20
Break Up of More Complicated Nuclei, 86-96
Break-Up of the Relativistic Deuteron and the Verification of Scale Invariant Properties of Its Wave Function, 74-85
Charged Lepton on Nucleus, The Process 1A → 1'(A – 1) X, Deep Inelastic Scattering of, 107-113
Collisions. The Process aA → 1^+1^-(A – 1)X, Lepton Pair Production in Hadron-Nucleus, 119-122
Composite Systems and Multiquark States in Nuclei, Deep Inelastic Form Factors of, 41-70
Composite Systems, Relativistic Elastic Form Factors and Scattering Amplitudes for, 21-40
Constituent Interchange Mechanism, 37-40
Construction of Tensor $W_{\mu\nu}$, 42-45
Deep Inelastic Form Factors of Composite Systems and Multiquark States in Nuclei, 41-70
Deep Inelastic Lepton-Nucleus Scattering with the Fragment Nucleus in the Final State, 107-136
Deep Inelastic Scattering of Charged Lepton on Nucleus, The Process 1A → 1'(A – 1) X, 107-113
Deep Inelastic Scattering of the Neutrino (Antineutrino) on the Nucleus. The Process νA → 1(A – 1)X, 114-118
Deuteron and the Verification of Scale Invariant Properties of Its Wave Function, Break-Up of the Relativistic, 74-85
Deuteron Relativistic Wave Function, Scale-Invariant Parametrization of the, 71-73
Dimensional Analysis and the Principle of Automodelity, 123-128
Elastic Form Factors and Scattering Amplitudes for Composite Systems, Relativistic, 21-40
Elastic Form Factors in the Impulse Approximation, 25-26
Electromagnetic Interaction, Lowest Order in the, 48-53
EMC-Effect, Quark Degrees of Freedom in Nuclei and, 57-70
Equation for the Many-Body Bound State Wave Function, 15-20
Equation for the Scattering Amplitude and Relation to the Equation in the Infinite Momentum Frame, 13-14
Equation for the Two-Body Bound State Wave Function, 9-12
Equation in the Infinite Momentum Frame, Equation for the Scattering Amplitude and Relation to the, 13-14
Formalism for Bound States, Light Front, 9-20
Formulation of the Method, 21
Freedom in Nuclei and EMC-Effect, Quark Degrees of, 57-70
Gluon Exchange in the Intermediate State, 35-36
Hadron-Nucleus Collisions. The Process aA → 1^+1^-(A – 1)X, Lepton Pair Production in, 119-122
^{3}He Nuclei, Scaling Properties of the Relativistic Wave Functions of ^{4}He and, 97-106
^{4}He and ^{3}He Nuclei, Scaling Properties of the Relativistic Wave Functions of, 97-106
High Energy Nuclei and the Problem of Relativization of Nuclear Wave Functions, Processes Involving, 71-106

Impulse Approximation, Elastic Form Factors in the, 25-26
Inelastic Form Factors of Composite Systems and Multiquark States in Nuclei, Deep, 41-70
Inelastic Lepton-Nucleus Scattering with the Fragment Nucleus in the Final State, Deep, 107-136
Inelastic Scattering of Charged Lepton on Nucleus, The Process 1A → 1'(A – 1) X, Deep, 107-113
Inelastic Scattering of the Neutrino (Antineutrino) on the Nucleus. The Process νA → 1(A – 1)X, Deep, 114-118
Infinite Momentum Frame, Equation for the Scattering Amplitude and Relation to the Equation in the, 13-14
Interchange Mechanism, Constituent, 37-40
Intermediate State, Gluon Exchange in the, 35-36
Lepton on Nucleus, The Process 1A → 1'(A – 1) X, Deep Inelastic Scattering of Charged, 107-113
Lepton Pair Production in Hadron-Nucleus Collisions. The Process aA → 1^+1^-(A – 1)X, 119-122
Lepton-Nucleus Scattering with the Fragment Nucleus in the Final State, Deep Inelastic, 107-136
Light Front Formalism for Bound States, 9-20
Lowest Order in the Electromagnetic Interaction, 48-53
Many-Body Bound State Wave Function, Equation for the, 15-20
Many-Body System, Relativistic Form Factors for the, 27-31
Model Parametrizations of Wave Functions, 54-56
Multiquark States in Nuclei, Deep Inelastic Form Factors of Composite Systems and, 41-70
Neutrino (Antineutrino) on the Nucleus. The Process νA → 1(A – 1)X, Deep Inelastic Scattering of the, 114-118
Nuclear Wave Functions, Processes Involving High Energy Nuclei and the Problem of Relativization of, 71-106
Nuclei and EMC-Effect, Quark Degrees of Freedom in, 57-70
Nuclei, Break Up of More Complicated, 86-96
Nuclei, Deep Inelastic Form Factors of Composite Systems and Multiquark States in, 41-70
Nucleon Structure Functions, Relations Between the Nucleus and, 129-136
Nucleus and Nucleon Structure Functions, Relations Between the, 129-136
Nucleus in the Final State, Deep Inelastic Lepton-Nucleus Scattering with the Fragment, 107-136
Nucleus, The Process 1A → 1'(A – 1) X, Deep Inelastic Scattering of Charged Lepton on, 107-113
Nucleus. The Process νA → 1(A – 1)X, Deep Inelastic Scattering of the Neutrino (Antineutrino) on the, 114-118
Pair Production in Hadron-Nucleus Collisions. The Process aA → 1^+1^-(A – 1)X, Lepton, 119-122
Parametrization of the Deuteron Relativistic Wave Function, Scale-Invariant, 71-73
Processes Involving High Energy Nuclei and the Problem of Relativization of Nuclear Wave Functions, 71-106
Quark Degrees of Freedom in Nuclei and EMC-Effect, 57-70
Relations Between the Nucleus and Nucleon Structure Functions, 129-136
Relativistic Composite Systems, Scattering of, 32-34
Relativistic Deuteron and the Verification of Scale Invariant Properties of Its Wave Function, Break-Up of the, 74-85
Relativistic Elastic Form Factors and Scattering Amplitudes for Composite Systems, 21-40
Relativistic Form Factors for the Many-Body System, 27-31
Relativistic Wave Function, Scale-Invariant Parametrization of the Deuteron, 71-73
Relativistic Wave Functions of ^{4}He and ^{3}He Nuclei, Scaling Properties of the, 97-106

Relativization of Nuclear Wave Functions, Processes Involving High Energy Nuclei and the Problem of, 71-106
Scale Invariant Properties of Its Wave Function, Break-Up of the Relativistic Deuteron and the Verification of, 74-85
Scale-Invariant Parametrization of the Deuteron Relativistic Wave Function, 71-73
Scaling Properties of the Relativistic Wave Functions of ^{4}He and ^{3}He Nuclei, 97-106
Scattering Amplitude and Relation to the Equation in the Infinite Momentum Frame, Equation for the, 13-14
Scattering Amplitudes for Composite Systems, Relativistic Elastic Form Factors and, 21-40
Scattering of Charged Lepton on Nucleus, The Process 1A → 1'(A – 1) X, Deep Inelastic, 107-113
Scattering of Relativistic Composite Systems, 32-34
Scattering of the Neutrino (Antineutrino) on the Nucleus. The Process νA → 1(A – 1)X, Deep Inelastic, 114-118
Scattering with the Fragment Nucleus in the Final State, Deep Inelastic Lepton-Nucleus, 107-136
Tensor $W_{\mu\nu}$, Construction of, 42-45
Two-Body Bound State Wave Function, Equation for the, 9-12
Verification of Scale Invariant Properties of Its Wave Function, Break-Up of the Relativistic Deuteron and the, 74-85
Wave Function, Break-Up of the Relativistic Deuteron and the Verification of Scale Invariant Properties of Its, 74-85
Wave Function, Equation for the Many-Body Bound State, 15-20
Wave Function, Equation for the Two-Body Bound State, 9-12
Wave Functions, Model Parametrizations of, 54-56